Ivory

By the same author

Gem and Ornamental Materials of Organic Origin

Ivory

Maggie Campbell Pedersen
FGA, ABIPP

NAG Press
an imprint of Robert Hale • London

To Gustav and Jasper
with my love

May there always be elephants in the wild in your world

First published in Great Britain 2015

ISBN 978-0-7198-0053-5

NAG Press
Clerkenwell House
Clerkenwell Green
London EC1R 0HT

NAG Press is an imprint of Robert Hale Limited
www.halebooks.com

A catalogue record for this book is available from the British Library

2 4 6 8 10 9 7 5 3 1

Design by Dave Jones
Printed in China

Contents

Foreword

If you are reading this you are the owner, or at least reader, of a much-awaited book by Maggie Campbell Pedersen, Fellow of the Gemmological Association of Great Britain, founder of the Organic Gems online information service and acknowledged world expert on organic gem materials. Organic gems, which include such animal or plant-derived materials as pearls, amber, jet and, of course, ivory, are very much a specialist subject. With Maggie we have not just the expert of experts, but one who gathers and disseminates information with a rare passion. I know this from personal experience of her writings, conversations with her and hearing her lecture to both general and specialist audiences.

In this book she focuses on one class of organic material – ivory – and all aspects of its use. In a world where plastics permeate our daily life, and landfills, it is difficult to imagine life before they existed, when ivory in its various forms could be an ideal, albeit costly, answer to a craftsman's or manufacturer's prayer. This is the story this book tells; it is a detailed and comprehensive study of the nature and properties of ivories and a history of their use over many millennia around the world. It covers all aspects, from Byzantine triptychs to false teeth, identification to conservation, microstructures to dating.

For those who instantly associate ivory with elephants, it is important to realize that there is a bewildering variety of animals that provide types of ivory, from walruses to whales. Elephant ivory may be the best known, but it now stirs a variety of emotions. Slaughtering elephants to produce piano keys strikes most people today as horrific, and most reasonable people abhor the illicit ivory trade. Against such a background, a new book extolling the beauty of art in ivory might well be condemned as tasteless. However, this is not such a book; it is a much-needed and scholarly study of all ivory types and their characteristics. Whether you are a customs officer or curator, a handbook like this is essential to help you distinguish between the many types of ivory and identify a plethora of substitutes and imitations. For the reader less in need of practical instructions, the author provides a history of ivory use and the ivory trade from Africa to the Arctic, uniquely interweaving science and history to present an erudite and readable narrative.

For my seventh birthday I was given my first book on archaeology and, having been born into a jewellery family, my interests meshed and for the best part of sixty years my library on jewellery history, materials and techniques has grown. Over that entire period the only book that was stolen from me was Maggie's 2004 *Gem and Ornamental Materials of Organic Origin.* That seems as good a yardstick as any to measure the extent to which Maggie's expertise in her subject is valued. Now, ten years later, she has produced a book on ivory which is also destined to become the standard work on the subject. I will keep my copy under lock and key.

Jack Ogden, Jewellery Historian and Consultant, London

Acknowledgements

When it was first suggested that I write a book on ivory I leapt at the chance as ivory is a fascinating subject with so many and varied aspects. In fact I had no idea of the task that lay ahead, because it is also a huge subject.

As is always the case, at the conception of an idea one thinks one knows it all, but it soon becomes apparent just how many gaps there are in one's knowledge.

I am therefore greatly indebted to the many people who have given freely and willingly of their time and knowledge in order to help me write this book. In one or two cases I have approached people whom I have never met but whose name I have stumbled across during my research, and found they have been willing to help. The quest for knowledge and the checking of facts has not only been fascinating; it has also been a heart-warming experience.

Nonetheless it is inevitable that mistakes creep into a book of this sort. Any mistakes here are entirely my own.

In my research I have read countless books and papers, visited numerous museums at home and abroad, worked briefly as a volunteer at an elephant sanctuary in Thailand, and consulted so many people along the way that it would be impossible to name every one of them. Special mention should be made, though, of the following, all of whom have added to the content of the book in one way or another. Some have been kind enough to proof-read sections of the book; others have lent or given items to be photographed; all of them have listened with patience and given excellent advice and encouragement.

Therefore I would like to thank (in no particular order) the following: the many curators and conservators at the Victoria and Albert Museum in London, including Joanna Whalley, Senior Metalwork Conservator; Alan Derbyshire, Head of the Paper, Book and Paintings Conservation Section; Sofia Marques, Sculpture Conservator; Nigel Bamforth, Senior Furniture Conservator; Dr Marjorie Trusted, Senior Sculpture Curator; Ming Wilson, Senior Curator, Asian Department; Dr Julia Hutt, Curator, Asian Department (Japan); Mariam Rosser-Owen, Curator, Middle East; Susan Stronge, Senior Curator, Asian Department; and Dr Paul Williamson, Keeper of Sculpture, Metalwork, Ceramics and Glass.

Also Dr Hans-Christian Gulløv, Senior Researcher, The Danish National Museum's Centre for Greenland Research; Fiona Sheales, formerly Curator of African Collections, The British Museum; Viviane Baeke, Senior Curator, Royal Museum for Central Africa, Tervuren, Belgium; Dr Terry Drayman-Weisser, Director of Conservation and Technical Research, Walters Art Museum, Baltimore, USA; Dr Joachim Meyer, Senior Curator, The David Collection, Copenhagen, Denmark; Peter Kristiansen, Senior Curator, The Danish Royal Collections, Rosenborg and Amalienborg Castles, Denmark; Dr Pierre Ickowitz, Director, Dieppe Castle Museum, France; Dr Barbara Simon Deutsches Elfenbeinmuseum,

Erbach; and Dr Richard Sabin, Senior Curator, The Natural History Museum, London.

Elise Skalwold, Senior Gemmological Curator, Cornell University, USA; Prof. Edward A. Alpers, Research Professor, History Department, UCLA, USA; Bear Williams, President, Stone Group Laboratories, Missouri, USA; Prof. Thomas Higham, Deputy Director, Oxford Radiocarbon Accelerator Unit, University of Oxford; Dr Jack Ogden, Jewellery Historian and Consultant, London; Alan Jobbins, President Emeritus, Gemmological Association of Great Britain; Julia Lucas, Master Ivory Carver, Germany; and Jane Tukarski, Scrimshander, Annapolis MD, USA. Also various people at the Earthwatch Institute, UK; the Gemmological Association of Great Britain; and the Golden Triangle Asian Elephant Foundation, Thailand.

Extra thanks must go to a few people who have supported me with great patience throughout the task. They include Caroline Armstrong, who has listened endlessly and proof-read; the Late Sonia Fraser who also listened and gave great encouragement; William R. (Bobby) Mann (co-founder of the International Ivory Society, USA) who has always been there to answer questions; my editors at NAG Press, Nikki Edwards, who must at times have wondered whether I would ever finish the book, and Lavinia Porter, who nursed me through the final stages; and, of course, my two lovely and supportive daughters, Louise and Kathrine.

To all of you, and to all the others who are not mentioned here, my heartfelt thanks. I could not have written this without your help.

Maggie Campbell Pedersen

Photo Credits

All the photographs in this book were taken by the author, with the exception of the following, which were kindly lent by museums (which retain copyright):

1.1, 1.48 & 1.51 © The Trustees of the Natural History Museum, London
1.27 © Ingo Arndt/Minden Pictures/FLPA
1.41 © Flip Nicklin/Minden Pictures/FLPA
1.46, 2.39, 2.40 & 3.15 © The Royal Collections, Rosenborg Castle, Copenhagen, Denmark
1.53 © Matthew Maran/The Trustees of the Natural History Museum, London
2.41, 2.52, 3.7, 3.9, 3.10, 3.13, 3.16, 3.27, 3.30 & 3.44 © The Victoria and Albert Museum, London
2.46 & 3.42 By kind permission of Jane Tukarski, USA
3.1 By kind permission of *Look and Learn*
3.2, 3.22, 3.23 & front cover image © The Trustees of the British Museum. All rights reserved.
3.3 & 3.12 © The National Museum of Denmark
3.8 © Pernille Klemp/The David Collection, Copenhagen, Denmark
3.11 & 3.34 © The National Museums Scotland
3.14 © Musée du Louvre, Paris, France
3.18 © The British Dental Association Museum, London
3.40 © The Auckland War Memorial Museum, New Zealand

Many items were kindly lent to the author for photography, including the following:

3.37 By kind permission of The Australian Museum, Sydney
1.15, 1.22 & 1.57 By kind permission of The Gemmological Association of Great Britain
1.5, 1.36, 1.44, 1.45, 1.55, 2.14, 2.15, 2.16 & 2.19 By kind permission of William R. (Bobby) Mann
1.49 & 3.28 By kind permission of Simon Rexworthy
3.41 By kind permission of John Wilkinson

Introduction

Ivory is a unique substance, unlike anything else. It is beautiful, and it is so tactile that it has been described as sensual. It can be worked in many different ways and put to many different uses. Throughout the history of its use – which goes back 40,000 years or so – in most parts of the world it has been highly prized, on a par with gold and gems. It has been collected as war booty, and it has funded wars. Over the ages the economies of several countries have been supported by the trade in ivory.

It is a material that humans are prepared to kill for, and to be killed for. Ivory is almost an addiction – one that has cost the lives of thousands of people, and of millions of animals.

Ivory, and the animals that bear it, have fascinated philosophers for thousands of years. Aristotle wrote about elephants over 2,000 years ago and much of what he wrote was correct, although he was wrong when he wrote that the tusks of males grow pointing downwards while those of females point upwards.

In the first century AD Pliny the Elder also studied elephants, and declared that 'Ivory is excellent for images of the gods', and that it was 'the most expensive material from those animals that breathe'. He also noted that it was unfortunate for elephants that they carried tusks, as this led to their demise.

Through the study of ivory we can learn large chunks of world history. In the Arctic ivory was a precious resource, used for utilitarian objects. In other parts of the world it has denoted wealth and status. In some areas it was in the sole domain of the Church or religious bodies, while in other areas only kings and rulers could use or own it. The famous throne of King Solomon, which is mentioned in the Bible, was made of ivory, as have been the thrones of several kings and queens since.

Where it was not locally available, ivory was imported. If a country ran out of its own supply of elephants or hippopotamuses because they had all been killed for their ivory, more was imported from elsewhere. The trade in ivory has crossed the globe from north to south, east to west, and back again for millennia and has been connected to the trade in spices, silks, gold, exotic animals, and, of course, slaves.

Ivory has been used for everything from supports for dwellings to the most intricate and beautiful carved works of art. It has reflected both mankind's love of beauty, and mankind's greed.

Today, at last, we are learning the true cost of our addiction. If elephant poaching cannot be stopped there will be no more viable elephant herds in the wild in twenty years' time. Walruses and hippos are now also in danger. Whales – which have been hunted for oil and for meat, and from which ivory was a by-product – are already protected but some countries want to lift the bans.

I have written about ivory for two reasons. First because I, too, have fallen under its spell, but secondly – and more importantly – because I believe that we need to know what it is. There is a surprising lack of knowledge about ivory and it is frequently misidentified at auctions or by dealers in antiques, and it is misrepresented by unscrupulous sellers both at home and abroad.

So what is this material? Why do we love it so much, and from where does it come? And how can we recognize it? With this book I have endeavoured to answer at least some of the questions. While there are numerous books that cover single aspects of ivory, I have attempted here to give an overview of all the different aspects, from the history and its various uses, to its identification and trade bans.

The text is divided into three sections. The first concerns the sources of ivory, i.e. the various animals. It deals with some of the history relevant to each animal, the animal's present conservation status, and a detailed account of how to identify each type of ivory. The second section covers fakes and simulants, various forms of testing, and care. Also covered is an explanation of the many ways of working ivory. In the third section I have looked at the uses of ivory through the ages, in various parts of the world, and have attempted to give an idea of 'Why?' as well as 'What and When?' Finally, there are two appendices of quick reference for identification.

Splitting the book into sections proved difficult, and there is some inevitable overlap. The illustrations are chosen primarily to help to explain a point and so do not always illustrate the most magnificent example. Fine examples can be found in many museums, worldwide, and I would urge readers to go and see them, the better to understand this beautiful material.

From a gemmologist's point of view, I have virtually ignored all the usual so-called 'constants', i.e. the numbers which describe such things as specific gravity. In my opinion they are of little value when looking at ivory, as testing the material in order to ascertain the constants is difficult to do without risking damage. The best way of identifying ivory is by visual examination, with a little common sense and some background knowledge. I hope that this book can be of help.

Maggie Campbell Pedersen

Section 1

What and Where?

The ivory-bearing animals and their conservation status
Identification of the different ivories
Uses specific to each animal's ivory

1.1

Conservation

It is far from being a new phenomenon that we have hunted animals to extinction. Elephants and walruses enjoyed a much greater range two thousand years ago. For example, elephants could be found in China and throughout much of the Middle East. It is worth remembering, though, that concern for the animals' numbers, and for their welfare, is a relatively new concept.

There is mention, historically, of one or two attempts by individuals to bring the subject of declining numbers to the attention of people in authority, but their concern was more about lack of resources for trade than about animal welfare, and anyhow they were largely ignored.

There are huge cultural differences in mankind's attitudes worldwide. In the West we are inclined to pamper our pets and in some cases treat animals as our equals: in the East there is generally a totally different attitude to animals. In many parts they are looked upon as existing for our convenience, while in Hindu areas cows are venerated for religious reasons, and in Buddhist areas the souls of animals are respected on a par with human souls. In underdeveloped countries it is no surprise that people regard animals as commodities, or as a huge and unnecessary nuisance when, for instance, an elephant tramples meagre crops.

The idea of conservation for the sake of the animals has only been around for about one hundred years, and to begin with it was simply to protect the environment and had nothing to do with animals. Organized protection of animals only began in earnest in the mid-twentieth century. Until then our love of decorative materials (such as ivory), which were derived from animals, had cost the lives of millions of creatures over several millennia, wiping out whole populations and, in some cases, whole species, until finally various organizations were founded to monitor and control our actions.

There was one exception. In Paris in 1905 a man by the name of Perrier, who was the Director of the Natural History Museum, was appalled by the slaughter of elephants in Africa, and rallied support from various quarters to start the Society of Friends of the Elephant. Knowing that it would not work if their aim was simply to spare elephant lives, he tried to sell the idea by suggesting that elephants could be put to much better use than simply as suppliers of ivory, and that they could be tamed and used as working animals, as was the custom in the East. However, it took another eighty-five years before a ban on ivory came into force.

CITES, the Convention on International Trade in Endangered Species of Wild Fauna and Flora, is the best-known of all the organizations. The name covers both the organizing body and an international agreement between governments. Its aim is to ensure that international trade in specimens of wild animals and plants does not threaten their survival. Concerns such are habitat loss do not come directly within its remit, although they are taken into account in its decisions. Their concern is international trade that contributes to the possible extinction of a species. Local problems are the responsibility of local conservation organizations and governments.

The idea for CITES was mooted by IUCN (*see below*) in 1963, but it did not come into full force until 1975. Any state can join CITES by making a formal declaration to adhere to and be bound by the Convention. Membership is voluntary. At the time of writing there are 180 signatories to the Convention, all agreeing to abide by the rules on international trade set out by CITES. In some rare instances a signatory can be allowed an exception or a moderation to a specific rule. Each member state must make its own domestic legislation to ensure that CITES rules are implemented at a national level.

CITES members meet every two years to agree how to classify the various species. The classifications spell out the trade bans according to the degree of protection the species need. There are three levels, called Appendices. Briefly, the classifications are as follows:

- **Appendix I** Species threatened with extinction. It provides the greatest level of protection, and includes bans on international trade. Exceptions can be made when the purpose of import is not commercial, i.e. for scientific research.
- **Appendix II** Species that are currently not threatened with extinction, but that may become so unless trade is controlled. International trade may be authorized for specimens in this category by the granting of an export permit or a re-export certificate.
- **Appendix III** Species which are being 'watched', or for which a member

country has sought help from other members in controlling trade. International trade in these species is allowed but only with the appropriate documentation and permits.

The International Union for Conservation of Nature (IUCN) was founded in France in October 1948. It publishes its own Red List of Threatened Species. Described as 'A neutral forum for governments, NGOs, scientists, business and local communities to find practical solutions to conservation and development challenges', has offices and volunteers worldwide. Its members are organizations or institutions. IUCN was instrumental in creating CITES, and it attends and contributes to all CITES meetings. IUCN undertakes research, surveys, and produces reports.

TRAFFIC was established in 1976 as a specialist group of the IUCN Species Survival Commission and works with WWF (The World Wide Fund for Nature) and CITES. TRAFFIC is the wildlife trade monitoring network – the 'wildlife watchdog'.

There are various intergovernmental organizations. For example, the International Whaling Commission (IWC) is tasked with the global conservation of whales and the management of whaling. Whales make transoceanic migrations and need global protection to survive. IWC was set up under the International Convention for the Regulation of Whaling, signed in 1946. IWC brought in the complete ban on commercial whaling in 1986.

Many other conservation organizations also work tirelessly in their efforts to protect animals and the environment. They carry out their own research, monitor situations, and advise governments. A few organizations combine tourism with conservation, and participants can pay to work with endangered animals – including elephants and whales – for a couple of weeks, their donations going towards conservation while they themselves get a chance to interact with the wild animals.

With regard to selling or purchasing items made from ivory, each country has its own laws for domestic trade, which is not affected by the CITES international agreement. These domestic laws vary from country to country. For example, the USA has various stringent laws on sales across state lines within the country.

Generally in Europe trade for commercial gain – which includes buying or selling at auction – is permitted for objects over a certain age. The age is determined by the trade bans, which are sometimes backdated. For example, the CITES ban on trade in all elephant parts including ivory was finally made law in October 1989 and came into force three months later, but was backdated so that an ivory item could only be sold if it was worked before 1947. The UK takes the law even further, and it is not permissible to sell pre-1947 ivory if it has been reworked after that date. The USA now had a total ban on the import of African ivory, no matter what its age, and permits only the export of African ivory items that are over one hundred years old.

Ivory carvings can be made from Arctic ivories by the local, native people (the Inuit, Yupik or Chukchi), as they are permitted to catch an annual quota of animals and work and sell their ivory as part of their subsistence agreement. The carvings can be purchased by others, but must have the appropriate certification. With this certification, permission may be granted to export or import a carving for own use, but not with the intention of reselling it, which would place it under the ban on selling ivory from later than 1947 for commercial gain.

Ivory which stems from the auctions organized by CITES in 1997 and 2008, when stockpiled ivory from four African countries was sold, can be purchased in China. The auctions were agreed on the condition that the material was for domestic use only. It cannot be taken out of China – even for own use and with no intention of resale – and imported into another country. In China there is a system of certification that accompanies any fresh ivory sold. Larger items must be accompanied by a photograph. Of course, such certificates can be 'forgotten' and re-used, and are anyhow quite easy to forge. It is an established fact that much more fresh ivory is sold in China than could possibly have come from the auctions sanctioned by CITES. The Chinese authorities come down very heavily on anyone found smuggling ivory into the country, and hand out huge fines and prison sentences of ten to fifteen years. Even buying ivory known to be illegal can result in a ten-year prison sentence. Yet much is overlooked and the illegal trade continues.

Many countries with populations of endangered species have set up systems to try to prevent poaching. Game reserves and wildlife sanctuaries employ armed rangers, but the areas they try to protect are vast and it is an impossible task to watch all the animals, all the time. When it concerns migratory animals such as whales the matter of protecting them is even more complicated and reliance is put on countries to keep their word and, where applicable, to keep to their quotas.

Poaching today is no longer the preserve of impoverished local people who see the opportunity to earn a couple of years' wages or more from the tusks of a single elephant, sold to a middle man. Nowadays it is well-organized crime, in some instances using helicopters and heavy firepower, and mostly organized by foreigners with access to the country through legitimate business – for example, factories – through which a few corrupt individuals can smuggle ivory out of the country.

In some areas the borders to two countries with different conservation goals are very difficult to police, and the animals may wander across them of their own volition. For instance, between Burma and Thailand some of the terrain makes patrolling impossible. In parts, the two countries are separated by rivers, and it is not unknown for elephants to swim across. Thus it can be difficult to know to which country the elephants 'belong'.

War presents a further danger to the protection of the various species. Examples of this can be seen in some African countries where all attempts at protecting elephants and other animals have been abandoned as large portions of the countries become no-go areas.

The sudden and complete ban on selling ivory goods internationally hit many honest craftsmen very hard. In places like Singapore, which had had a lively market for locally carved ivory, there were suddenly no customers as almost all their trade was international. Today only one or two local craftsmen continue to work the ivory that they already had in stock, as they can only sell to the local population, and there is a limit to how much the local people wish to buy. Some of the craftsmen moved abroad, but most simply stopped carving.

Similarly, many ivory carvers elsewhere had no ivory to work, and no customers, and were forced to learn new carving techniques on other materials. They could start to use bone instead, but customers felt that bone was very inferior – a belief still held today, although very beautiful carvings can be made in bone. Another alternative was to use mammoth ivory, and it is to this that some craftsmen turned. Others had no alternative but to switch trade altogether, or to close.

In spite of all the concern, the regulations and the bans, illegal ivory is still being sold, both in the East and in the West. In Europe and America it can be found in some antique shops or at antique fairs, where it is labelled 'pre-ban' or 'antique' although there is no proof of the fact and no documentation to back up the assertion.

Another area where the law is constantly flouted is online auction sites. In some countries (for example, the United Kingdom) the sale of all ivory – including legal antiques – was banned by the largest online auction sites a few years ago, in an effort to control the problem. However, the battle with unscrupulous sellers continues as ways are found to get around the ban. The favourite method is to be inventive with the description, so terms such as 'faux ivory', 'ivory-coloured', and 'bone ivory' are used for items that are clearly genuine ivory.

Some of the lots for sale are antique and pre-ban, but a large percentage may well be new, and – even online – prices can be extremely high. The price paid by the elephants, walruses and whales, meanwhile, is the highest of all.

1.2

Mammoths and Mastodons

Mammoths and mastodons are not the predecessors of elephants, although they all descend from the same order of animals called proboscideans, which have trunks and tusks. Proboscideans appeared about ten million years after dinosaurs became extinct, and included such animals as gomphotheriums, and stegadons. It is only recently that enough DNA has been extracted from a mammoth to obtain a complete mitochondrial sequence and to be able to compare the different species in more depth.

Mammoths were larger than mastodons, although the many species of mammoth varied greatly in size. The American mammoths were huge, while elsewhere some isolated, island herds developed as dwarf mammoths. The ivory that is dug out of the permafrost comes from woolly mammoths. The males measured about 3 m at the shoulder, the females a little less.

Mammoths were the descendants of other animals of vaguely similar shape. The first mammoths emerged in Africa and spread north, appearing in Europe about 1.5 million years ago. From there they spread further, reaching Siberia and finally North America. As they moved to different parts of the world they acclimatized, some of them eventually developing into the woolly mammoths of the frozen north.

The first stories of these creatures which were found in the ice reached Europe in the seventeenth century, when the tales of enormous animals living in the frozen wastes were probably based on a mixture of mammoth remains and living walruses.

Although mammoth ivory had long been traded by the local populations, it was not until early in the twentieth century that the first scientific expeditions to discover these monsters took place. The only way to reach the area was by sled. It took many months to travel there and back, and any soft tissue on the remains had rotted by the time they were brought back to Europe, leaving just bones, teeth and tusks. Today helicopters and ground-penetrating radar can be used to find remains in the permafrost, which is enabling far better research. Even so there have only been a few finds of complete mammoth skeletons, and even fewer still have soft tissue attached.

Owing to the deterioration of the mammoth carcasses found in warmer climes, mammoth ivory comes almost exclusively from the woolly mammoths that lived in the north. The slow loss of water into the icy sediment aids the preservation of woolly mammoths, and they have, in many cases, remained intact for thousands of years in the permafrost. The ivory from these long-dead animals can be in very good condition, though it is estimated that over 50 per cent of the ivory found in the permafrost is not of suitable quality for carving. Ivory from warmer areas – for example, mammoths found in England – is cracked and brittle and it would be highly unlikely to find any that could be carved. The value of these tusks is purely scientific.

Mammoths died out over a period of many thousands of years, until they were finally extinct. It has been discovered that one species of mammoth in Siberia died out about 40,000 years ago, while the last mammoths lived until only 4,000 years ago on a small Siberian island. There was a reported sighting of a mammoth in Siberia at the beginning of the twentieth century, but the country consists mostly of vast expanses of wilderness, so it would be difficult to prove this one way or the other. The idea seems, however, rather fanciful.

Mastodons were about the same size as today's Asian elephant, and had a straighter back than the mammoth. The mastodon's tusks were also shorter and less curved. They lived until about 11,000 years ago, and inhabited North America from Alaska to Mexico, with concentrations around the Great Lakes and the Eastern Seaboard. Their predecessors were also mastodons and lived in North America, Europe and Asia, but the most common one – and the one from which it is most likely to find tusks – was the last: the American mastodon, *Mammut americanum*.

There have been finds of mastodons in other parts of the world. For example, a spectacular find was made in Greece in 2007, where the remains of a prehistoric mastodon were unearthed with a complete pair of tusks, one of which measured 5 m long, and the other 4 m. Other mastodons have been found in England, Germany and the Netherlands. All the European mastodons are thought to have died out about five million years ago.

Teeth and Tusks

MAMMOTHS

Mammoths, like today's elephants, had two forms of teeth: tusks and molars. Mammoth molars were similar to those of elephants, in that they had a 'running shoe pattern' – that is, ridges running across the surface of the tooth. They also had six pairs of molars in both the upper and lower jaws (making a total of twenty-four). The molars were developed gradually throughout the life of the animal. They were not all present at the same time, but appeared in pairs which were slowly worn down by the action of eating. The worn teeth were gradually pushed forward by new teeth emerging from behind and replacing them, as the old teeth disintegrated and fell out (*see also* Section 1.3 Elephants: Teeth and Tusks).

Fig. 1.1 Woolly mammoth molar showing 'running shoe pattern' on the exposed surface.

As with elephant molars, the dentine and enamel occurs in a series of folds, so that the tooth, cut perpendicular to its surface, will appear banded unless the cut exactly follows the folds of the dentine and enamel. There is no structure of intersecting arcs in molars.

Mammoth tusks grew throughout the life of the animal, and adult mammoth tusks were considerably larger than those of elephants. They had a spiralling trajectory (although they seldom completed more than half a turn) so that in an old animal they often crossed.

The outside of a mammoth tusk did not have any enamel, but when a tusk is found today the cementum layer present is hardened by age and usually dark brown. It often shows signs of cracking and the exterior can be rough to the touch. This outer layer is called 'the bark', and is usually removed before any work is done on the ivory, although it can also be left on as part of the design of a piece.

In cross-section mammoth tusks are round. They also display an intersecting arc pattern (also known as Schreger Lines) similar to that of the elephant, but there is one important difference. In mammoth tusks the arcs are at an acute angle (that is, less than 90 degrees) while in elephant ivory they form obtuse angles. This is most pronounced at the edge of the tusk, as the angles narrow closer to the centre, making it less easy to judge the angle of the elephant ivory's intersecting arcs. In mammoth ivory meanwhile, the angles become very narrow. This is of great importance when identifying proboscidean ivories, as it means that ivory displaying such an acute angle can be legally traded. At

Fig. 1.2 Cross-section of a small mammoth tusk, showing the dentine with a central dark spot, Owen Lines (*see text*), intersecting arcs of less than 90 degrees, and the cementum. Diameter 8.5 cm.

Fig. 1.3 Pair of mammoth ivory cabochons, showing clear structure of dentine and cementum, and the demarcation between them. Diameter 2 cm.

the centre of the section is a small black dot. It is the remains of the pulp cavity which has closed and disappeared as the animal grew, the dentine converged and the tusks got longer.

The longitudinal cracks which occur naturally in a tusk may penetrate the dentine inside. Further, mammoth ivory frequently cracks along the 'Owen Lines', which are concentric growth lines in the tusk. As a point of weakness in the structure, cracks appear here rather than haphazardly throughout the tusk.

► **Fig. 1.4** Piece of mammoth ivory, showing the bark and cracks along Owen Lines.

▼ **Fig. 1.5** Piece of mammoth ivory (the tip of the tusk is to the left) showing cone-in-cone structure and cracks along Owen Lines.

MASTODONS

Mastodon molars are different from elephant or mammoth molars as their surface is a series of cusps, rather than a ridged pattern. The cusps are in two parallel rows, and each has a root growing into the jaw, but they are joined together, forming one long tooth. In cross-section they have a different appearance from mammoth molars as the enamel, dentine and cementum does not occur in folds, but as in, for example, human molars, with enamel on the outside of the cusps, cementum underneath the enamel, and a core of dentine. Each part has a pulp cavity ending in the root.

Fig. 1.6 American mastodon molar, showing cusps on the exposed surface.

Some mastodon molars are extremely worn on the surface, so the cusps have virtually disappeared.

Because of the structure of the molars they would not be suitable for carving in the way of mammoth molars, which have a structure that is a decorative feature.

Mastodon tusks were not as large as those of the mammoths, and they were much less curved. In cross-section the tusks display the typical proboscidean intersecting arc pattern (Schreger Lines), but the angle of the arcs resembles elephant ivory rather than mammoth ivory, in that, viewed near the outer edge of the tusk, the angles are more than 90 degrees. Generally speaking, the structure is a little finer than that of an elephant tusk.

See also Appendix I, which lists the main identifying features of the various ivories.

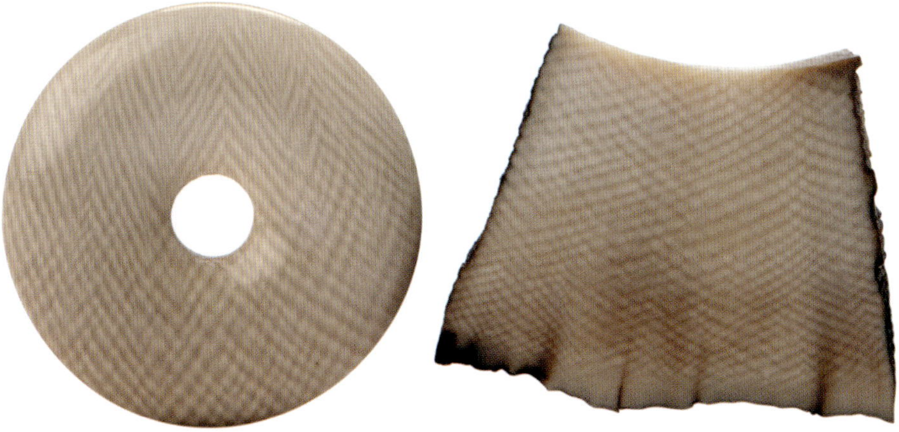

Fig. 1.7 Mammoth disc showing acute angles of intersecting arcs (left), and piece of mastodon ivory showing obtuse angles of arcs (right). Both show cross-section of the tusk. The disc measures 2.5 cm diameter.

TUSK ANOMALIES

Just as with any other teeth and tusks, the tusks of these animals could also grow in an irregular fashion. There are examples of mammoth ivory which display areas of extraordinary patterns of nodular material, with little sign of intersecting arc structure. The nodules can sometimes be discoloured so that they appear to have dark outlines. Other samples can be found where an area of the normal tusk has disappeared altogether, and only a coarse material consisting entirely of nodules remains. Such material is clearly not of much use for fine carvings, but can be put to very good use in imaginative creations in which the anomaly is integral to the design.

Uses and Beliefs

Mammoths were vital to humans in the Ice Age, as they not only provided food, but their bones and tusks provided building materials, and their bones could be burned as fuel. Once alight they burned for a long time – much longer than wood.

The name 'mammoth' is taken from two Russian words meaning 'earth' and 'burrower'. Originally the appearance of mammoth remains in Siberia, slowly emerging from the melting silt, caused immense consternation. They were thought to be underground-dwelling monsters emerging from the earth, that died when they surfaced as they could not tolerate air. Not knowing what they might do between surfacing and dying, the locals decided that they were to be avoided at all costs. The possibility of trade finally overcame these fears, however, and mammoths were being dug up and their ivory sold to markets in the Middle and Far East long before Russia invaded Siberia in 1582. Old Chinese writings mention the import of mammoth ivory for carving in 400 BC, and by the tenth century ivory was being sent both to other parts of Asia and to Europe. Nonetheless, when a collection of mammoth bones was dug up in Switzerland in 1577, nobody knew what they were, and there was discussion as to whether or not they should be given a Christian burial. However, the idea was abandoned when a doctor examined the bones and identified them as coming from an eighteen-foot tall giant. As giants were considered to be heathens, the bones could not receive a Christian burial.

Russia took over the trade and exported quantities of mammoth ivory in the seventeenth century. By the early 1900s Siberia was an accepted source of ivory for the billiard ball and piano key trade, yet it was only at this stage that people in Europe really found out what the material was and where it was coming from, and scientific research into the species began.

The research and trade in mammoth tusks virtually stopped under Communist Russia, but in the 1990s, following the freeing up of the Russian economy and the ban on international trade in elephant ivory, it started again. Today mammoth ivory fetches very high prices, and 90 per cent of the material is exported to China. A small amount is retained in Russia for domestic use.

In Siberia the tusks may only be extracted by people who have permits, but it is very hard to police such a large area and there are plenty of people who are willing to take great risks in order to find and sell the precious commodity. Permits are also required in Alaska, where there is greater control and tusks can only be traded by the indigenous population or people with mineral rights in the area. As the Alaskan tusks can be of good quality and therefore of monetary value, there is not a problem with preserving them, but in many other parts of the world, where remains are found in conjunction with mining or quarrying of some sort, repeated stopping of production to allow excavation by archaeologists proves too costly, so finds are often covered up and ignored.

The American mastodon ivory from Alaska can be carved and worked in the same way as mammoth ivory, but it is much rarer and very little is found that is of workable quality.

The first sign of a mastodon was a large tooth found in 1705, which was presumed to have come from a giant. When the bones of the animal were subsequently found, it became apparent that it did not have a very large human shape, but still nobody knew what it was, and

Fig. 1.8 Chinese carvings in mammoth ivory, 5 cm high, made in 2005.

Fig. 1.9 Slice of Alaskan mammoth tusk with cracks filled in with decorative materials, and made into a pendant.

whether it walked on all fours or upright. It was called the *Incognitum*. It took a hundred years to discover exactly what it was and how the bones fitted together. Even then, when the assembled skeleton was exhibited, the tusks had been inserted into the skull upside down, so that they pointed downwards. In 1806 the animal was finally given the name 'mastodon' by a French anatomist, Georges Cuvier, who decided that the cusps on the molars resembled miniature breasts and so took the words from the Greek: *mastos* for' breast' and *odont* for 'tooth'.

As may have already been inferred, the names 'mammoth' and 'mastodon' can be slightly misleading and have at times seemed to be almost interchangeable. It is quite possible that some ivory labelled as 'mammoth' may in fact derive from a mastodon; indeed the name 'mammoth' has tended to be used generically for all very large extinct animal finds, just as the name 'mastodon' has been used to refer to any of the several large, extinct proboscideans of the genus *Mammut* which look like elephants, but which have molars of a different structure.

To further complicate matters, the classifications of proboscideans are frequently revised as relationships within the order can be controversial. Some animals formerly described as mastodons are now officially classified as mammoths.

It could be said that, except for scientific purposes, it is immaterial whether ivory comes from a mammoth or a mastodon as both are extinct and therefore neither is controlled by international trade bans. However, there is one problem. While mammoth ivory can be distinguished from elephant ivory by the internal structure of the tusk, the American mastodon ivory closely resembles elephant ivory (*see* Teeth and Tusks, *above*). This can lead to complications in identification and trade. Fortunately, there is very little mastodon ivory on the market.

A question remains as to whether or not there should be free trade in ivory from extinct species. It is not a renewable source, and one day there may be no more tusks left to dig out of the permafrost. It seems likely that this will not happen for a very long time as it is estimated that there could be millions of mammoths still buried.

Nonetheless, there are other problems connected to the collection of ivory from extinct species. First, the extraction of tusks from the earth while leaving the rest of the animal *in situ* distorts the scientific picture. Much is lost that could be of value to researchers, who use the knowledge gained to put together a picture of the world in times past and the animals that populated it. For example, some of the well-preserved mammoths have died and been frozen so fast that their stomach contents could be analysed. The different plant types ingested give us a better picture of the climate at the time.

Another potential problem is that new ivory is sometimes sold as mammoth ivory in order to get around trade bans, but there have been few cases of this so far and it would seem to be isolated incidents only.

An emotive question is whether the use of mammoth ivory is helping to save the elephant by presenting an alternative source of ivory, or whether it is in fact encouraging the possession of ivory to the extent that it is actually encouraging elephant poaching and slaughter. Again, so far the evidence is inconclusive, so there are no plans at present to curb the trade in mammoth ivory. The general consensus is that, as mammoths and mastodons are already extinct and there is no risk of endangering them, their ivory can be traded freely.

Although it is legal to take mammoth or mastodon ivory across most international borders (there are one or two countries that do not permit it), it is worth remembering that many people would not be able to recognize the difference between these ivories and elephant ivory, especially in the case of mastodon ivory, so it would always be advisable to carry appropriate documentation if attempting import or export.

1.3

Elephants

Elephants are the largest of the living land mammals. Today, there are two species, the African elephant, *Loxodonta Africana*, and the Asian elephant, *Elephas maximus* (also known as the Indian elephant).

Originally there were many more species of elephants, but these have slowly died out, leaving just the African and Asian species. The former has two sub-species: the savannah (or bush) elephant, *L. africana africana*, and the forest elephant, *L. Africana cyclotis*. (There is some controversy as to whether African elephants are in fact two separate species rather than sub-species, in which case the forest elephant should be called *Loxodonta cyclotis*, but at present they are treated as one species.)

Fig. 1.10 African elephant in the Ngorongoro Crater, Tanzania, using its tusks to strip bark from trees.

The Asian elephant has three sub-species, although this fact is often ignored and they are all regarded as the same. The sub-species are: *Elephas maximus maximus*, which lives in Sri Lanka, *Elephas maximus indicus*, which lives in India, and *Elephas maximus sumatranus*, which lives in Sumatra.

It has been suggested that *Elephas maximus borneensis* from Sabah and Sarawak on Borneo should also be named as a sub-species, but this has yet to be decided. The Indian sub-species has one pair of ribs fewer, but otherwise they are all much more similar than the two African sub-species, with only very slight differences in size and colouring. As there can be variations within each sub-species it would be hard to tell them apart by sight alone.

Fig. 1.11 Male Asian elephant with small tusks.

Elephants are the only remaining members of the order Proboscidea. The name Proboscidea derives from the Greek and its literal translation is 'front feeder' (though today it is accepted as meaning 'possessing a trunk'). Elephants' trunks are their noses, which are amazingly versatile limbs that the animal uses to rip up grass or bamboo, to drink (by siphoning up water and squirting this into its mouth), to smell and feel, to touch and caress, to lift and strike, and also to pick up very small objects (for example, an elephant is capable of picking up a fruit by its stalk). The trunk has over forty thousand muscles.

Elephants are fascinating animals in many respects. Socially, the female elephants live in matriarchal groups, while the males live mostly solitary lives, although there are some bachelor herds.

Elephants are not aggressive animals, but, because of their immense size and power, they can be dangerous if angered. Periodically, male elephants 'come into musth', which is a state of heightened sexual drive. At this stage they can be easily angered and are constantly ready to fight, although their natural opponents are other bull elephants. Female elephants are normally placid; nonetheless, they are always protective of their family groups, especially the young.

Elephants cannot jump but must always step over uneven ground. They cannot turn their heads in isolation, but must turn their whole body. Nevertheless, they are very agile creatures and can manoeuvre such obstacles as narrow flights of steps without being able to see where they are going. They are also surprisingly silent on their feet for a creature of such size.

It is believed that elephants favour one side more than the other with their tusks, just as humans are right- or left-handed. This is indicated by one tusk generally being more worn that the other. African elephants are larger than Asian elephants. A large male can be well over 3 m tall at the shoulder, and weigh up to 6 tonnes. The females are slightly smaller. Among other physical differences between the species, African elephants are known for their huge ears and single, unpronounced dome on the top of the head. The trunk has two pointed areas known as 'fingers' at the tip to help in picking up objects.

Asian bull elephants are seldom more than 3 m at the shoulder and weigh less that 5 tonnes. The females are correspondingly smaller as well.

All Asian elephants have two distinct domes on the top of their heads, and smaller ears than the African elephant. They have only one 'finger' at the end of their trunk, but it is as versatile as that of their African cousin. The females do not grow large tusks, but have small tusks called 'tushes', which are easily visible when the mouth is open. Only the tip may be glimpsed when the mouth is closed.

Fig. 1.12 Young female Asian elephant bathing and drinking. Note the small 'tushes'.

Among the males, 'tuskers' – that is tusk-bearing elephants – are becoming less numerous. Some elephants are genetically disposed not to grow tusks. As more and more of the tuskers have been killed for their ivory, allowing the non-tusk bearing males to reproduce, the tusk strain is becoming weaker. Although sad to acknowledge, this is undoubtedly going to be good for the survival of the Asian elephant, and may be one of the reasons why Sri Lanka now has a denser population of elephants than other Asian countries. It is reckoned that only 2 per cent of the Sri Lankan elephant population have tusks.

Borneo is home to the so-called pygmy elephant. It is not a very small creature, but instead of standing just under 3 m at the shoulder like the other Asian bull elephants, the pygmy bulls are about 2.5 m at the shoulder. They also have tusks that are smaller and straighter.

Status and Conservation

Elephants are not descended from mammoths or mastodons but are their cousins, both species having branched from the same order of animals: the proboscideans.

The African elephants alive today live only in sub-Saharan Africa, while the Asian elephants can be found in India, and many of the Far Eastern countries. In times past elephants roamed much wider areas. Elephant remains have been discovered near the ruins of Babylon in today's Iraq, which date back to 1800 BC. It is thought that they became extinct in Mesopotamia in about 850 BC, and there are records of elephant hunts in the first millennium BC in the area that is today's Syria, which accounted for large numbers of elephant deaths.

Elephants are thought to have lived in China until as late as AD 200. All these animals are believed to have been Asian, or a sub-species of Asian elephants. African elephants – or possibly a sub-species of African elephants – were found north of the Sahara in Africa until Roman times.

Today we tend to think of Africa as the main habitat of elephants, and the Asian ones are sometimes forgotten. But it was the Asian elephants that played a big role in history, while the African ones were considered more as a source of meat, hide and ivory.

Elephants cannot be domesticated, but they can be tamed, and humans have been catching and taming elephants for over 4,000 years. Although the Asian elephants have been more used throughout history than the African, it is possible to tame and train both species. All over Asia elephants have been used in battle. Starting in India the custom spread across South-East Asia, and elephants were used in some areas until late in the nineteenth century.

Elephants made up part of the army, much as horses have done, but elephants were larger and far more frightening. It is thought that they were used for war purposes in China as long ago as 1200 BC. They were trained for combat, and their role was generally to charge the enemy. Understandably, the opposition would scatter in disarray in the face of a line of charging elephants. In North Africa the Carthaginians used them for battle, the most famous example being the trek made by Hannibal, who took an army of soldiers and elephants across the Pyrenees and the Alps in 218 BC to attack Rome.

Around the Mediterranean the elephant tamed and used for various purposes was the North African Forest Elephant. This sub-species died out as the herds were depleted until they were no longer sustainable, and they became extinct in Roman times. One of the causes of the diminishing numbers of elephants from northern Africa was their frequent use in the arena for fights against gladiators or against other animals. The fights were stopped in AD 523 largely because the attitude of the spectators changed and they felt pity for the elephants.

By the twentieth century elephants were no longer being used for war, but were used extensively for heavy-duty work such as logging, hauling and building. In some countries in South-East Asia logging is completely banned, and more are realizing that it must be curbed. Although the logging bans are good news for the planet, this sudden stop means that there are a lot of tamed elephants with no work. For example, when logging was banned in Thailand in 1989, over 3,000 elephants were made redundant overnight. No work meant no food for the mahouts and their families, and no food for the elephants either. Many of them were forced to take work wherever they could get it, and hundreds of elephants and their mahouts are now in elephant camps where they work, sometimes in very poor conditions, for the tourist trade, while others tout for business on the streets.

Elephants have been used in processions and ceremonies for thousands of years, especially in Asia. The custom continues today. They have also been used as performance animals in circuses – another custom that continues today, though it is far less frequent than before. The only other ivory-bearing animal that is consistently being used for the entertainment of humans is the orca, also known as the killer whale (*see* Section 1.8 Orca), which is used for displays at marine parks.

Historically, the ivory in Asia came partly from wild elephants, and partly from those in captivity which had died of natural causes. In most parts of Asia the wholesale slaughter of elephants for their ivory started later than it did in Africa, as in Asia elephants were more important as workers. In Africa the massive slaughter of elephants for ivory began with the first traders.

Elephants do not have any natural predators, though a young calf can occasionally be taken if left unattended by its mother. This is a rare occurrence as elephant females are fiercely protective, and if the calf's mother is not there to protect it, other females in the herd will do so. The one, real threat to elephants is mankind. An expanding human population and the subsequent destruction of the elephant's natural habitat has meant that elephants can no longer roam freely. In Africa most elephants are corralled in game reserves and sanctuaries where over-population can occur, resulting in too little food, or the destruction of the habitat of another species such as the black rhino. There are ongoing discussions about the pros and cons of culling in these cases, and most game parks in Africa permit culling when deemed necessary.

The biggest problem that elephants face today is without doubt that of illegal poaching. There are various methods used: shooting the elephants, catching them in barbaric traps, or poisoning them. All risk injuring the creature without killing it, and possibly leaving it to a painful and lingering death. Traps and poison are also a danger to other animals.

Ivory has always been sought after, and with today's expanding markets in the East and the new-found wealth in those countries, the demand for ivory is higher than ever. Poaching was originally carried out by poor local people who could see no reason not to kill an elephant and earn enough to feed their families for several months. The animal had probably been helping itself to their meagre crops anyhow. Today the expansion of Eastern commerce into Africa has opened up new possibilities, and many of the poachers caught are of Oriental origin. In some countries war and strife make it impossible to protect the wildlife, and in those countries the ivory is poached and sold to fund wars. African elephant numbers will soon be so reduced that deaths will outnumber births, making extinction inevitable. At the present rate of killing, the African elephant will be beyond saving within twenty years.

Wherever possible poaching is combated. In various countries game wardens are permitted to shoot to kill the poachers. But where there is money to be made quickly, high risks will be taken, and without very large and costly armies of wardens, it is utopia to think that all poachers will be caught. Those that are caught are dealt with harshly, and the ivory is, of course, confiscated. But this gives rise to another problem: what to do with the tusks from culled or poached elephants? Some countries in Africa have now obtained permission from CITES to hold auctions of their stockpiled ivory. So far permission has been granted twice to hold auctions: the first in 1997 and the second in 2008. The conditions for granting permission to sell the ivory were very strict and the ivory was sold exclusively to China and Japan for their domestic use. Further regulations concerning the onward sale of the material or anything crafted from it had to be put in place, and the money raised from the auctions was to be used for elephant conservation.

Many people are against auctions as they believe these encourage the passion for and use of ivory, thus fuelling the illegal trade. Following the first auction poaching figures actually decreased for a while. However, there was no immediate change after the second auction, and numbers of elephants killed by poachers has increased year on year since then, till today poaching is at an all-time high.

At present elephants or any part thereof are listed on CITES Appendix I by all but four African countries – those that have sold stockpiles. These countries are Namibia, Botswana, South Africa and Zimbabwe. In some African countries it is permitted to hunt elephants for sport, for which very large sums of money are taken with the idea that this will aid conservation. There are strict quotas set by CITES for the numbers permitted to be killed, as well as their ages and gender.

Today efforts are being made in China to re-introduce Asian elephants in Yunnan Province, but although they are in a nature reserve there, they are constantly coming into conflict with the local population. The future of the project is uncertain.

African elephants are listed as 'vulnerable' by IUCN, and Asian elephants as 'endangered'.

Teeth and Tusks

African elephant tusks tend to be larger than those of their Asian cousins, and the tusks of the forest elephant are straighter and narrower than those of the savannah elephant. Both males and females carry tusks, the females' tusks tending to be smaller than those of the males.

The ridges across the surface of African elephants' molars – the so-called 'running shoe pattern' – are diamond-shaped.

As already stated, only the male Asian elephants carry true, visible tusks. The females have small tusks called 'tushes', which are easily visible when the mouth is open, but only the tip can be glimpsed when the mouth is closed. Tushes do not have the same structure as elephant tusks, and the pattern does not appear as continuous curved lines that intersect, but as a mass of small, broken ones. There is also little or no pulp cavity. However, the chance of encountering it in a worked item is very remote, as tushes were only rarely used as ivory.

The 'running shoe pattern' on the surface of Asian elephants' molars appears as a series of slim, parallel ovals.

▲ **Fig. 1.13** The open mouth of an Asian elephant, with the molars in the top jaw just discernible. The molars have a 'running shoe pattern' of slim, parallel ovals.

Although it has often been stated that baby elephants are born with 'milk tusks' (that is, deciduous tusks), it seems that so few people can state that they have seen them – including people who work with elephants – that their existence has been queried. However, elephant calves do indeed have deciduous tusks called 'tushes' that can grow up to 5 cm in length. They do not erupt from the jaw so they do not fall out, and are therefore never seen. They remain in the jaw until they disintegrate and are reabsorbed by the baby elephant. The permanent tusks start to be visible when the elephant is about two years of age.

The shape of elephants' teeth is very different to most other living mammals, as is the manner in which the teeth grow. In total an elephant has twenty-six teeth: two modified incisors, which protrude from the upper jaw and are called tusks, and twenty-four molars. The molars grow in rotation during the elephant's life. At any one time they have one good molar in each side of the upper and lower jaws which they use for chewing and grinding their food. At the same time the next set of four molars is growing from the back of the mouth and slowly pushing the good ones forward. As they become worn through use they gradually disintegrate until the last vestiges are gone, and the next set of new molars take their place. The first set, which is usually replaced when the elephant is about two to three years old, is small. Subsequent sets are larger as the animal grows until it reaches maturity. The last set appears when the elephant is forty to forty-five years old. This set must last for the remainder of the elephant's life. If it wears out completely and becomes unusable, the elephant dies of starvation.

The molars have an unusual structure. They are made up of folds of dentine and enamel, held together by cementum, the structure giving them their bands of creamy material sandwiched with off-white, translucent material. This is very apparent when seen in longitudinal section. The surface of the molars is covered in cementum and the upper surface is ridged. The ridges become more pronounced with use as the cementum wears away and the harder enamel underneath wears better than the exposed dentine.

The eating motion is not up and down, as for a human, but is more of a

circular movement with the teeth in the upper jaw grinding forward over the surface of those in the lower jaw.

Elephant tusks grow throughout the life of the elephant, and some of the largest have attained a length of over 3 m. They are, as mentioned, modified incisors, embedded deep in the upper jaw. It is useful to understand the shape of an elephant's tusks when identifying ivory, as the typical structure may be visible only from odd angles, or be otherwise hidden. This is because, in order to avoid waste of such a precious material, a tusk may be cut into sections that are not totally straightforward (*see* Fig. 2.36).

▲ **Fig. 1.14** Section of elephant molar, showing folds of enamel and dentine, covered with cementum.

The pulp cavity extends about a third of the length of the tusk. The whole tusk has a thin coating of cementum and may have a small amount of enamel on the tip, though this is often worn away.

The dentine nearer the pulp cavity is softer than the rest, and is avoided in fine carvings as it is said to be less durable and may also develop a slightly darker colour with age, and the cone-in-cone structure becomes more visible. In larger carvings it is usually hidden at the back of the item.

The dentine close to the outer edge of the tusk is also avoided where possible as cutting through the cementum can give a slight colour variation which shows on rounded areas of carving as an ellipse. If an item is carved entirely from the edge of the tusk, it can be difficult to see any sign of an intersecting arc pattern, and the item may instead display very irregular, long patches which vary in depth of colour. This can be very misleading as, on occasion, it resembles plastic more than ivory in appearance.

▼ **Fig. 1.15** An unworked elephant tusk.

◄ **Fig. 1.16** Interior of a box used for cosmetics. The base has been carved from material near to the pulp cavity and has absorbed more stain, making the cone-in-cone structure visible.

▼ **Fig. 1.17** Paper knife, carved into the cementum, giving a false impression of the uneven patterning and colouring that can be encountered in plastics.

► **Fig. 1.18** Elephant ivory beads, carved very close to and into the cementum.

▲ **Fig. 1.19** Detail of a bead showing demarcation between dentine and cementum.

The outer area of dentine is drier than that at the inside of the tusk and is prone to hairline cracks which run longitudinally in the tusk. All in all only about 60 per cent of a tusk gives ivory of the best quality.

The cone-in-cone growth structure (caused by the pulp cavity continually laying down dentine) is not readily visible in a fresh tusk, but will show more clearly in an old piece of ivory, especially if it has dried out and cracked as the cracks tend to follow the cones. However, when viewed in longitudinal orientation the structure of irregularly parallel, slightly wavy lines can be seen. These are the result of the orientation of the dentinal tubules (*see* Section 2.1 What is Ivory?).

In cross-section it is possible to see the diagnostic proboscidean structure of intersecting arcs (also called Schreger Lines, named after the German anatomist Bernhard Gottlob Schreger who first described them in 1800). This pattern is also caused by the orientation of the dentinal tubules, which form two helices curving in opposite directions, giving the decussated arc pattern. In elephant ivory

▲ **Fig. 1.20** Back of a hand mirror, cut to obtain optimal size from the tusk, showing cementum at the edges on either side.

► **Fig. 1.21** Elephant ivory box, the sides showing the structure of the tusk in longitudinal section.

◄ **Fig. 1.22** Two slices taken in cross-section from an elephant tusk, showing the intersecting arc structure (*left*) from around pulp cavity and (*right*) showing clear demarcation between cementum and dentine.

the angles formed are obtuse (that is, more than 90 degrees). At the outside edge of the tusk they are wider - up to 120 degrees – and are more pronounced. Nearer to the centre of the tusk the angles become narrower and less discernible. (Mammoth tusks display intersecting arcs with acute angles.)

The tusks are round or almost round in cross-section, with a small black dot at the centre which is the remnant of the pulp cavity. The intersecting arc pattern spreads out from the centre. However, not all tusks grow completely round, nor do they all have the dot exactly in the centre, which can result in a slightly lop-sided formation of arc pattern and some arcs being of a more acute angle than normal.

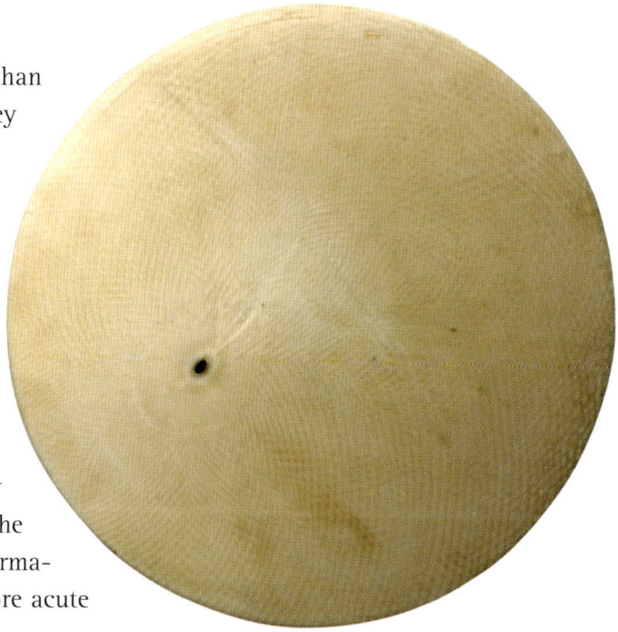

▲ **Fig. 1.23** Base of polished box cut from cross-section of tusk. Note that it is slightly off-centre, which affects the shape of the intersecting arcs; also, the dark spot is not central.

The pattern of concentric growth rings caused by the cone-in-cone growth structure, called the 'Owen Lines', is not usually very noticeable in new elephant ivory, though it can be seen. If ivory dries out cracks can appear along these lines.

There has long been a belief that, depending on its origins, elephant ivories differ, with some having a finer grain, some being harder, some whiter, some having a more yellow hue, and so forth. One reason given for the different qualities and tints in fresh elephant ivory is diet. Thus African elephants grazing in the savannah were said to produce creamy white ivory, those in the forest a more brown ivory, and those feeding largely on bamboo produce ivory with a faint pinkish hue. While this sounds very logical it has not been proved. Further, Asian elephants tend to eat a lot of bamboo, yet their ivory is white. Nonetheless, it is generally accepted that fresh Asian elephant ivory is slightly whiter than African, and that ivory from the forest elephant is different to that of the savannah elephants in Africa, and is closer in texture and hardness to Asian elephant ivory.

Ivory carvings can become more mellow in colour with the passage of time, and some elephant ivory items become more yellow while others remain white. There are theories that Asian elephant ivory retains its pale colour better than African ivory, but again there are conflicting reports, with some records noting that it is the other way round. However, it is generally accepted that, especially in a carved item, it is not possible to determine by sight where the ivory came from, nor from which extant species of elephant – even if the ivory is fresh.

USUAL IDENTIFYING FEATURES OF ELEPHANT IVORY

- Size of item carved.
- Curve of tusk possibly visible in shape of a large carved item.
- Round or off-round cross-section shape of tusk.
- Intersecting arc pattern (Schreger Lines) in cross-section of tusk, with approximately 120-degree angles at outside edge. These can appear 'stretched' or distorted when the item is carved at an angle to the axis of the tusk.
- Irregular parallel lines in longitudinal section of tusk.
- Possible ellipses on rounded areas of carving.

Fig. 1.24 Back of a hand mirror showing the intersecting arc pattern from an unusual angle, giving it a distorted appearance.

- Possible cracks in longitudinal section, or following the Owen Lines.

Note: to see the structural identifying features in elephant ivory, it is necessary to turn the item during inspection, as the features are not apparent from every angle.

See also Appendix I, which lists the main identifying features of the various ivories.

▼ **Fig. 1.25** Damaged dentine that has repaired itself

TUSK ANOMALIES

Tusks on a living animal can be damaged yet continue to grow. On occasion, when cutting up a tusk after the death of an elephant, some foreign object has been found inside the dentine – for example, a bullet – left there when an elephant had been inaccurately shot and not died as a result. Around the bullet the dentine had continued to grow, covering it and 'healing' the wound. The dentine produced was of a very disorganized, globular pattern, in no way resembling the intersecting arc pattern, though sometimes resembling the secondary dentine of walrus ivory. This form of dentine was known by traders and carvers as 'pith'; large globules were termed 'beans'.

Similarly, broken tusks can grow to heal the broken area, and new dentine covers any open pulp cavity. Although the tusk is deformed in structure, and the broken tusk will never grow sufficiently to match the unbroken one in length, it can continue to function.

Another form of elephant dentine anomaly is the rounded, calcified dentine masses, which can form in the pulp and are known as 'elephant pearls'. They may display growth layers on the surface, but they do not have any pattern of intersecting arcs (Schreger Lines). As they lack any real identifying features they may need to be tested by DNA in a laboratory to confirm their origin. 'Elephant pearls' were mentioned as being one of the nine sacred gems in ancient Vedic texts, and were revered by Buddhists as being almost magical. They did not know how or where the 'pearls' were formed, but the common belief was that they came from inside the temple of an elephant. The concretions are very rare and expensive, and consequently are often faked, sometimes using carved elephant molars.

It is possible that some anomalies are genetic and run in families of elephants,

▼ **Fig. 1.26** Damaged dentine that has repaired itself, producing globular material somewhat resembling walrus ivory secondary dentine in structure. The intersecting arc pattern is visible at the end of the piece.

as seems to be the case in some pods of toothed whales. Elephant tusks with anomalies have in the past been of great value to odontologists, as they enabled the study of dental lesions – especially in the pulp – in very large format. They could also be used by carvers, giving special attention to the anomalies, and using them to effect as part of the carving.

Uses and Beliefs

Elephant molars are not very often seen in worked form. As they are made up mostly of alternating layers of very hard enamel and much softer dentine, the material is very difficult to carve. This structure also gives them a unique appearance. When carved into items such as scent bottles, the pattern from the structure is part of the decoration. It is possible to dye the material which gives a striking effect as only the dentine layers absorb dye, while the enamel layers remain a neutral off-white colour.

The uses of elephant tusk ivory are very many and varied. It is the most versatile of all ivories because, with the exception of mammoth ivory, it is the largest and most solid form of the material, and while mammoth tusks may be larger, they often have discolouration and cracks. Elephant ivory has been used for everything from chair legs to toothpicks, and from the most exquisite jewellery to hip prostheses. Thanks to its shape, size and mass, it can be turned into large display cups or carved into large figures. It can be made into intricate, pierced work, softened and woven into fans, bent to resemble ships' sails, sliced paper-thin so that it is translucent, used for inlay or veneer, coloured, painted on, or itself burned to make black paint. It has also been thought to have talismanic and medicinal properties.

It is probably simpler to list the items for which elephant ivory was less suitable than other types of ivory, rather than to list its actual uses. One such example is as dentures, for which elephant ivory was considered too soft. There were also times when it was not readily available in northern Europe, such as in the Middle Ages, and an alternative was found in the form of walrus ivory. With extremely few exceptions elephant ivory has been the ivory of choice for the past 6,000 years or so, with only its availability, or the cost of the material as opposed to its cheaper rivals, deciding that it should not be used.

There are two items in particular which have historically been manufactured from ivory and which subsequently helped to decimate elephant populations: piano keys and billiard balls. Although it is only a small sliver of ivory that covers each piano key, so many pianos were being made that vast amounts of ivory were needed to satisfy demand. Finally, early plastic substitutes were found, but they were not popular with pianists because the 'feel' was not the same. Because it is hygroscopic (tends to absorb moisture), real ivory is less slippery to the touch than plastic, which gives more control to the pianist.

Although not a new game, billiards and other similar games became extremely popular in the late nineteenth and early twentieth centuries, and over the years hundreds of thousands of billiard balls were manufactured. Tusks of a certain size and width were chosen, to prevent wastage of material and time, and only three billiard balls could be made from a single 'ball tusk'. Further, billiard balls could easily warp if not stored at an even temperature, or they could crack, so they frequently had to be replaced.

Elephants have long been creatures beloved of children's book authors, and have been the subject of myth and magic for centuries. They feature in religions, for example, the elephant-headed Hindu god Ganesh, who is also worshipped by Buddhists. In India an old belief held that earthquakes involved elephants. Four of them were said to hold the world on their backs, while they stood on the back of a turtle. The turtle in turn stood on the back of a cobra. If any of the animals moved, it would cause the earth to shake.

1.4

Walrus

The walrus, *Odobenus rosmarus*, is a large, bulky mammal that lives in colonies in coastal areas of the Arctic and Sub-Arctic. The genus is divided into two sub-species: the Arctic walrus, *Odobenus rosmarus rosmarus*, lives in the Eastern Canadian Arctic and along the coasts of Greenland and Ellesmere Island, and the slightly larger Pacific walrus, *Odobenus rosmarus divergens,* inhabits the Chukchi and Bering Seas, from the north-eastern coast of Siberia to the north-western coast of Alaska. The Pacific walrus male can reach a length of 3 m, and usually weighs between 1 and 1.5 tonnes, while the females are slightly smaller. The Atlantic walrus is approximately 15 per cent smaller than the Pacific species.

Recently it has been suggested that there is a third sub-species, *Odobenus rosmarus laptevi*, which lives in a small area around the Laptev Sea (part of the Arctic Ocean near Siberia), but research so far indicates that this group shares many common traits with the Pacific walrus and should remain classified as such. Its slightly smaller size could be the result of its limited habitat.

Walruses look like very big seals with tusks and slightly mournful expressions. In fact, they are distantly related to seals as they, together with sea lions, are all pinnipeds (meaning 'fin-footed'). Walruses move much better in water than on land. Although they have flippers, they move through the water by sinuous movement of the whole body rather than by moving the flippers. On land they turn their rear flippers forward and use them to walk on all fours, though the gait is more of a shuffle than a walk.

Walruses have a circulatory system that can withstand extreme cold, and breathing abilities to allow them to dive to a depth of about 80 m to find the molluscs that live on the seabed. The females are very maternal and protective of their young.

▲ **Fig. 1.27** Pacific walrus bulls resting, Bristol Bay, Round Island, Alaska.

Status and Conservation

The fact that the walrus is another animal that has been hunted for its ivory is largely overshadowed by the plight of the elephant.

Walrus habitat used to extend much further south than it does today, and walruses could be encountered as far south as Newfoundland. Constant hunting has meant that instead of the millions that used to live in the world, walruses can now be counted in the thousands.

The hunting began in the Stone Age, as a subsistence source for local populations, but it was not until the ninth century when walruses were being hunted by the Russians and Norwegians for trade with countries further south that this became exploitation. By the eleventh century, walrus ivory was also being traded by the Norsemen in Greenland for the vital supplies without which they could not survive.

From about 1800 until the middle of the nineteenth century, Arctic walruses were hunted in their thousands as it became increasingly widely realized that, in addition to their tusks being solid ivory, their blubber made excellent lamp oil and their skins could be made into very strong rope. Additionally, in Russia,

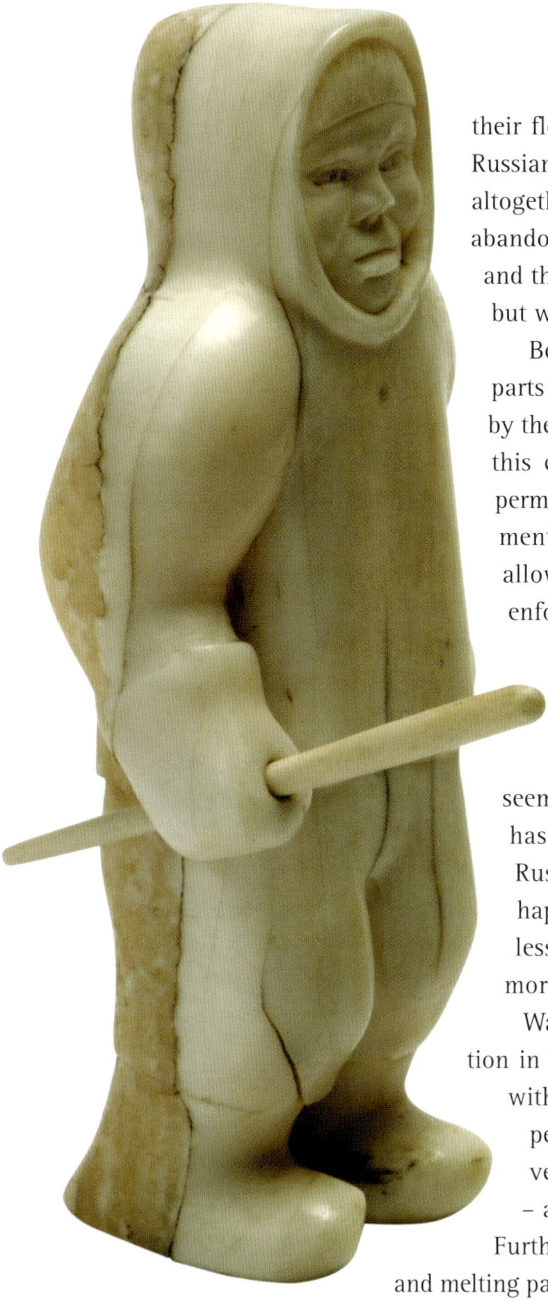

▲ **Fig. 1.28** Carving of an Inuit hunter, with primary dentine at front and secondary at rear. Made in Greenland in early twentieth century. Height 8 cm.

their flesh was used as a food source in fox fur farms. However, the Russian trade eased off in the latter part of the century, and stopped altogether with the Russian Revolution in 1917. The fox fur farms were abandoned, so walrus meat was no longer needed to feed the foxes, and the ivory carving industry also ceased to exist. It did start again, but was never as great as before.

Between 1931 and 1972 hunting bans came into force in other parts of the world, and today only subsistence hunting of walruses by the various indigenous peoples of the Arctic is permitted, although this can prove difficult to enforce. The local people are further permitted to work and sell the ivory themselves as this gives employment and brings in some much-needed money, but they are not allowed to sell unworked tusks. Again, this can be difficult to enforce, and it has been suggested that there is more native-carved ivory around than could possibly come from the annual quotas. It is also believed that there is a rising level of poaching being carried out.

However, while the Atlantic walrus is under threat, it seems that the Pacific walrus population has not only steadied, but has begun to recover. This is probably, in part, a result of the Russians having lost interest in the ivory at an earlier stage than happened in other parts of the world, but also because there is less pollution in those areas than in the North Atlantic, as it is more remote.

Walruses are very sensitive to any form of disturbance, be it pollution in the water or the presence of industry and the noise that goes with it. Walrus blubber has been found to absorb heavy metals and pesticides. Added to this, the reproductive cycle of the walrus is very slow – the females give birth only every two to three years – and the animals only live to an age of about twenty-five years. Further, the reproductive cycle is affected by the warming climate and melting pack ice, and has recently been seen to be even slower than before.

Apart from humans, walruses have two natural predators: polar bears and orcas. The former seldom pose a very great risk, although they are known to target a calf if it becomes separated from its mother on the ice. Orcas are larger than walruses and can pose a real threat by working as a team and separating a few walruses from the rest of the group, so that they are much easier prey.

All walruses are listed on CITES Appendix III (*see* pages 16–17), which means that their populations are being watched as there is a known risk that they could become endangered.

Fig. 1.29 Unworked walrus tusk.

Teeth and Tusks

The walrus's most eye-catching feature is a pair of long and solid tusks, which are modified upper canine teeth. They grow downwards in a slight curve, and in the Pacific animals can reach a metre in length. The Arctic walrus tusks are usually smaller, at about 60 cm. Both male and female walruses carry tusks, but the females' tend to be both shorter and slimmer than those of the males.

The purpose of the tusks is to help the animal manoeuvre itself out of the water, and to hack the edges of holes in the ice to prevent them from freezing over. This is done by lying on the ice and using the tusks as ice picks. Another purpose could be to aid the animal when swimming near the sea bed, where it can trawl in murky depths in poor visibility. With only its tusks touching the surface and acting as runners, the animal avoids bumping its head. The reasoning for this theory is that the front of the tusks in an adult animal are well worn, suggesting that they have undergone constant abrasion.

The tusks have a small pulp cavity and are solid dentine for most of their length. They are tough and heavy, as is needed when used to help the animal drag its great weight onto an ice floe. There is seldom any enamel on a mature walrus's tusk, as this is usually worn away during the animal's life.

Unique to walrus tusks is the structure, which is made up of two different types of dentine. In cross-section the tusk is vaguely oval in shape, with irregular indentations. This can vary from animal to animal. The outer ring of cementum is thin, and inside this is a ring of dentine – the primary dentine – which appears as faint, finely banded, concentric lines. Inside the primary dentine is a mass made up of nodules or globular material, also described as 'looking like rice or tapioca pudding'. This is the secondary dentine. It usually has an almost identical colour to the primary dentine, but a slightly more

Fig. 1.30 Cross-sections of walrus tusks: (left) unworked walrus tusk slice showing primary and secondary dentine, and the cementum layer surrounding the dentine; (right) polished slice of fossilized walrus tusk, showing typical discolouration. Width of fossilized slice is 6 cm.

Fig. 1.31 Proximal end of two walrus tusks, and a slice taken from the proximal end of a third tusk (left). Note the 'irregular oval with indentations' shape and the secondary dentine just visible.

translucent appearance. The reason for the difference in structure is that the secondary dentine contains no dentinal tubules (*see* Section 2.1 What is Ivory?).

In a live walrus the proximal end of the secondary dentine is partially surrounded by pulp, so the pulp cavity is not totally conical as it is in an elephant tusk. If a tusk is viewed from the proximal end, the outside surface of the secondary dentine is knobbly and looks somewhat like chopped walnuts.

The secondary dentine is less suitable for carving than the outer, primary dentine. Although it takes a good polish, it cannot be carved in as fine detail and is said to be prone to crack and splinter in carving.

Apart from the tusks a walrus has several straight, peg-like teeth, none of which measure more than about 5 cm in length. An exact number for these teeth cannot be given as they vary somewhat, but a total of eighteen teeth might be considered normal. The peg teeth get flattened on the distal surface during the life of the animal by wear and tear. They have a thick layer of cementum covering the dentine. Seen in cross-section the cementum displays concentric lines and there is a clear demarcation line between cementum and dentine, and the secondary dentine is little more than a spot or a trace of globular material.

See also Appendix I, which lists the main identifying features of the various ivories.

▲ **Fig. 1.32** Base of an *okimono* depicting a Samurai warrior, with secondary dentine disguised by incised, stained lines. See also the whole figure, Fig. 3.35).

FOSSILIZED WALRUS IVORY

The term 'fossilized' suggests that at least some of the material has become mineralized and the organic part transformed into calcified material. However, in the case of fossilized walrus ivory, it is a misnomer as this has not happened, yet the terms used for the material are 'fossilized walrus ivory' or simply 'fossil walrus ivory'. Further, much of the ivory is not very old. Although some may be three thousand years old, much of it is only three hundred years old.

It is found along the coasts of Arctic Alaska and Canada, and by law may only be harvested by the local, indigenous people, who are permitted to sell it. Some washes up on the beaches but a lot is found by scuba-diving during the warmest period of the summer when portions of the ice melt.

◄ **Fig. 1.33** Three slices of fossilized walrus incisors, showing thick cementum, dentine, and a trace of secondary dentine at the centre.

Fig. 1.34 Fossilized walrus ivory scraps, showing typical discolouration.

Fig. 1.35 Frog carving in fossilized walrus ivory. Note the discolouration of the primary dentine and the clearly visible structure of the secondary dentine.

The old material looks much the same as fresh walrus ivory, though most of it is somewhat discoloured by absorption of minerals from the earth in which it has lain. Usually it takes on a darker brown tint, more pronounced on the outer edge, and possibly leaving the secondary dentine the original creamy colour. It can be polished to the same high lustre as fresh walrus ivory.

Fossilized walrus ivory is often cracked and broken. The discolouration is used to advantage in the making of jewellery, for which small pieces are ideal.

Theoretically, fossilized walrus ivory can be traded legally as it is from animals that are no longer under threat of extinction, but have been dead for a very long time. It is popular in the United States, where it is freely available, but it can prove difficult to export from there and to import into another country because its status is a somewhat grey area. In some countries it is regarded as deriving from an extant species, and is therefore covered by the same trade bans as walrus ivory.

Uses and Beliefs

As with other animals that lived in faraway lands and were only seen by a handful of people, the walrus's appearance was historically rather a mystery. Various descriptions suggested that it was a sea-dwelling, man-eating monster, or that it was a fish the size of an elephant. Drawings of the hapless creatures were also bizarre, usually showing their tusks growing upwards from the lower jaw instead of downwards from the upper jaw, and the bristles on their muzzles, called 'vibrissae', were thought to cover the whole body. In 1555 Olaus Magnus – a Swede living in Rome – published *The History of the Northern Peoples*, and illustrated a walrus as a rock-climbing beast with upward-pointing fangs.

The scientific name Odobenus comes from the Greek, and means 'tooth-walker'. The French call walrus *morse*, and that name is sometimes used in the English language to refer to the ivory. In fact, walrus ivory was usually labelled 'seahorse ivory' when it appeared at the ivory sales in London in the eighteenth to twentieth centuries.

To the Inuit and Eskimos, the walrus was food, fuel, skins for their kayaks, rainwear (made from the intestines), and tusks for building materials or harpoon points. The well-being of the walrus impacted directly on their own well-being, so their lives were intertwined. People and animals were all guarded by the spirits, with which walruses were deemed to be closely connected.

In the Middle Ages, at the time when walrus ivory carving in Europe was at its height owing to the paucity of elephant ivory, many very beautiful carvings were made for ecclesiastical purposes. Two of the most beautiful are the so-called Gunhild Cross in Denmark (*see* Fig. 3.12), and the Cloisters Cross, now in the Cloisters Museum (part of the Metropolitan Museum) in New York. Both are thought probably to date from the mid-twelfth century, and both are exquisitely worked, with great care taken only to use the primary dentine for the areas with intricate carving (*see* Section 2.5 Determining the Age and Origins of Ivory Carvings). Also made in mediaeval times, and even more famous than the two crosses, are the Lewis Chessmen (*see* Section 3.4 Europe).

▲ **Fig. 1.36** Dyed walrus ivory (secondary dentine only) imitating jade.

About seven hundred years later walrus ivory was commonly used for something totally different: dentures. With its secondary dentine, walrus ivory might seem a strange choice of material for dentures, but it was less expensive than hippo ivory, and elephant ivory was too soft, as well as being expensive. The dentures were carefully made with the teeth carved from the primary dentine, positioning the secondary dentine as the gums and plate.

Walrus had been the ivory of choice for sword and dagger handles made in the Ottoman Empire. It was considered to give a better grip than elephant ivory. Similarly, it was often used in later years in Europe for umbrella and stick handles.

Another use, though not a common one, was as a jade imitation. Only the secondary dentine was used, and it was dyed green. For many reasons it was not a very convincing fake: it was warmer to the touch than jade, it was lighter in weight than jade, and the nodular structure was not typical of jade.

1.5

Hippopotamus

The hippopotamus, *Hippopotamus amphibias*, or 'hippo' as it is usually called, lives in sub-Saharan Africa. It is one of the largest land animals (third after elephant and rhinoceros), and can grow to 1.65 m at the shoulder, and weigh up to 3,200 kg. The hippo's closest relatives are cetaceans – that is to say, whales and dolphins.

Hippopotamus amphibias is one of only two extant species of the family Hippopotomidae. The second is the pygmy hippopotamus (*Choeropsis liberiensis*), which lives only in a few remote forested areas in West Africa. It is an endangered species in the wild, but it thrives in captivity and is often seen in zoos. Standing at less than half the height of its cousin, it has relatively small teeth and it is not thought that these pygmies were ever used for their ivory. The near demise of the pygmy hippo is a result of loss of habitat.

Hippos are mostly nocturnal, and they are herbivorous. They spend their nights eating huge amounts of grass and other vegetation, and most of the daylight hours lying or standing in shallow water in ponds or rivers, where they can keep cool, their only movement being a waggling of the ears to expel water.

The name derives from the Greek and means 'river horse'. However, they are not good swimmers, but prefer, when possible, to walk, even if that means walking under water, where they are able to stay submerged for up to five minutes between breaths.

Despite their bulk hippos are fast on their short legs, and can outrun a human. They are considered one of the most dangerous animals in Africa as they are not only big, but also aggressive.

Status and Conservation

The habitat of the hippopotamus is wide-ranging, covering most of sub-Saharan Africa. Some thirty thousand years ago hippos populated the whole of Africa, and also Asia and Europe. They were still to be found in the countries bordering the Eastern Mediterranean as recently as the start of the eleventh century, and in Upper Egypt until the nineteenth century.

In the wild, hippos live to an age of about forty to fifty years. They are one of the few animals that can do well in captivity, where they live longer. They also breed well in captivity.

Because of their immense size, hippos do not fall prey to many other creatures, but calves can sometimes be taken by crocodiles, lions or hyenas. However, other dangers await the hippos, one of which is loss of access to water. Another danger is war. It is reckoned that up to 90 per cent of the hippo population of the Democratic Republic of the Congo was lost during the Second Congo War (which started in 1993 and was officially ended with a peace accord ten years later). Further, unregulated hunting and poaching of the animals continues in many African countries.

Hippos have always been killed for their meat, but – apart from the Ancient Egyptians whose main source of ivory was the hippopotamus – it is only in the last two hundred years or so that they have been killed for their large, solid teeth. This is because hippo teeth are covered in a layer of enamel which must be removed before working the ivory. The enamel is extremely hard to strip off, although the task became easier with the advent of a rotary grindstone, first used in the Middle Ages. Elephant and walrus ivory did not need as much preparation, so were preferred materials for the people working ivory at that time. They were also a more versatile shape, as the large hippo teeth are very curved.

Nevertheless, whenever elephant and walrus tusks have become more scarce or expensive, people have turned to hippo ivory. This has been especially true since the bans on elephant ivory came into force. Ethiopia and Tanzania, where good hippo populations survive, have permits to export raw hippo teeth, and large amounts are sent to China and Japan. Although rules and regulations vary from one country to another, it is still possible – with the appropriate documentation – to import hippo ivory into most countries.

In 2006 the IUCN listed hippopotamuses as 'Vulnerable'. It was estimated that the population had declined by up to 20 per cent on the preceding ten years. Since then it will have declined further. Hippo ivory is not yet banned. However, rules change and it may soon become one of the totally banned ivories.

▲ Fig. 1.37

Hippopotamus in a zoo. The large canine teeth in the lower jaw, of which two-thirds are embedded in the jawbone, are the main source of hippo ivory.

Teeth

Hippo teeth are not visible when the animal's mouth is shut. Because of the structure of the jaw, the mouth can open very wide. This gesture is misconstrued as a yawn when it is in fact one of threat.

When the hippo opens its mouth an extraordinary array of irregularly placed teeth is seen. The purpose of the rear teeth is for grinding up leaves and grass, but, while the front teeth may conceivably be used in the process of eating by helping to gather vegetation, their real purpose is for fighting. The mouth can open to almost 180 degrees and expose the teeth, so they make fearsome weapons in a fight. The incisors block an attack, while the canines can inflict serious damage.

At the back of the hippo's mouth are twenty-four molars, six on each side in the upper and lower jaws. They are square with a cusp at each quarter, and are used to grind food.

In the upper jaw are two slightly curved canines. They are covered in vertically ribbed enamel, except on the outer surface where they are worn from

▲ **Fig. 1.38** Unworked hippo lower canine.

grinding against the canines in the lower jaw, and the cementum is exposed. Through wear the dentine is also partially exposed and is worn away at an angle. In cross-section the teeth are best described as slightly kidney shaped. They have a pattern of very fine concentric lines, following the outline of the tooth. At the centre is the slightly curved TIZ (the tooth interstitial zone), which is caused by the dentine converging as the tooth grows and the pulp cavity is filled in. It is the equivalent of the dark spot at the centre of elephant ivory. In cross-section it appears as a dark line, or can resemble a crack. Because the TIZ is curved, at certain points when the tooth is cut in longitudinal section, both ends of the curve are seen, giving rise to two dark lines, where other ivories have only one.

The canines in the lower jaw are the big teeth from which most ivory is derived. They are very large, strongly curved and heavily ridged lengthways, solid for most of their length, and set very deep in the jaw so that only about a third of the tooth is visible above the gum line. They can measure 70 cm along the outside of the curve and weigh up to 3 kg. They have an enamel covering, except for the area that is worn away by grinding against the canines

▲ **Fig. 1.39** Hippo incisor (left) and hippo lower canine (right), both seen in cross-section. Note the very fine lines in the dentine, and in the incisor, the TIZ and the cementum layer. Width of incisor, 7 cm.

in the upper jaw. As with the upper canines, the exposed dentine is worn down at an angle, resulting in sharp-edged teeth.

In cross-section the lower canines are more triangular in shape than those in the upper jaw, with indentations caused by the vertical ridges on the teeth. The dentine is seen to be dense, and to display extremely finely packed concentric lines, following the outline of the tooth. There is a clear demarcation line between the dentine and the thin layer of cementum surrounding it, and at the centre of the tooth is the interstitial zone. As with the upper canines, cut longitudinally the TIZ sometimes appears as two dark lines.

Between the upper canines are four incisors which are peg-like, blunt (or broken), and may be rather short. They are also covered with vertically ribbed enamel.

In the lower jaw are two pairs of peg-like incisors, one longer pair of which inclines forward, while the other pair may appear as pegs, or may even remain as small stumps. The incisors in the lower jaw are almost round in cross-section, and display a fine structure of concentric lines. They also have a thin layer of cementum, but they have only a small cap of enamel and are less ridged than

▲ **Fig. 1.40** Detail of hippo incisor showing fine structure of lines. The rule to the left is in millimetres.

canines. A small dark spot may be visible at the centre of the tooth, which is a remnant of the pulp cavity.

The incisors and canines grow continuously throughout the life of the hippo, but as the animal eats the teeth in the upper jaw grind against those in the lower jaw, with the result that they are worn down. The molars are also worn down by the motion of eating, but as they do not grow continuously they can wear down to the extent that the animal is no longer able to eat, and it dies. If one of the canines is broken – for example, in a fight – the hippo is left with a big problem because the opposing canine continues to grow unhindered, which can make eating extremely difficult.

The identifying features of hippo ivory are not always easy to distinguish because the dentine is very dense. The very fine concentric lines and – where it is visible – the TIZ, are the best indications of ivory being from a hippopotamus.

See also Appendix I, which lists the main identifying features of the various ivories.

ANOMALIES

Not only can it be difficult to see the identifying features of hippo ivory; very occasionally there is a misleading anomaly which makes identification even more complicated.

It has been reported that hippo ivory can display a pattern near the edge of the tooth that is not dissimilar to the intersecting arcs of proboscidean ivory, at the same time as displaying the very fine lines that are characteristic of hippo ivory. However, it is an extremely rare occurrence and the likelihood of encountering such examples is not great. The reason for the anomaly is unknown.

Uses and Beliefs

Cave paintings from about 3,000 BC, found in Africa, depict the hippo as a very brave beast. It was popularly thought of as braver than a lion.

Hippos once lived all along the Nile, so the Ancient Egyptians used hippo ivory rather than elephant ivory. This is a little surprising as it must have been a very difficult material for them to work because they did not have efficient tools for removing the hard enamel coating. It is possible that they dissolved it away in acid, although it would have been necessary to have done this with great care in order not to damage the dentine underneath.

Once it became possible to work it with ease with the help of more modern tools, the advantages of hippo ivory were recognized in Europe. It is whiter than other ivories, slightly harder, denser, and more homogenous as it has less visible structure. It is also stain resistant. It was therefore ideal for the manufacture of dentures and other items that might otherwise stain. It became a popular medium for *netsuke* and *okimono* in Japan, and for carving in China, where large, intricate display carvings made a feature of the strong curve of the tooth.

With the ban on elephant and whale ivories, hippo became an alternative and today a lot is sent to China where it is worked into many different decorative items, including intricate carvings and fake scrimshaw (*see* 'Scrimshaw' in Section 2.8). Some is also worked in Africa for sale to tourists, even though special permits would be needed to export it (*see* also Fig. 3.26, modern carved hippo tooth).

In Western culture the hippo has been embraced and depicted as a big, fat, friendly creature. The rotund figure is seen to be amusing and endearing, and has featured in many cartoon films. One example can be seen in the Disney film *Fantasia* where the *corps de ballet* was made up of a line of tutu-wearing bipedal hippos. Soft, cuddly toys are made in the image of hippos, and children's books have many friendly hippo characters. This is a far cry from reality, as the hippo is an aggressive beast, solitary, quick to anger, and not at all cuddly.

Yet the Ancient Egyptians clearly did not feel that hippos were unpleasant animals. Tawaret, the Ancient Egyptian goddess of childbirth and fertility, who protected and helped women, was commonly depicted as a bipedal hippopotamus with some feline features. A hippo tooth was carved with images and used as a magic wand, probably to draw a circle around a mother and child to protect them.

1.6

Narwhal

The narwhal, *Monodon monoceros*, is a small Arctic whale, with a body that measures about 4 m from nose to tail. It very similar in size and shape to the beluga but it is much darker in colour, and the males have the amazing feature of a long, spiralling tusk. Both beluga and narwhal belong to the family Monodontidae.

The narwhal is one of the deepest divers of all whales, and can dive to a depth of 1,800 m. Five hundred years ago the tusk of the narwhal was believed to be unicorn horn.

Status and Conservation

The narwhal lives mostly in the northern Atlantic between Greenland and Baffin Island in Canada. There are a few pods of narwhals in and around Hudson Bay, and further along the Arctic Ocean towards Russia.

Narwhals are migratory. As the polar sea ice melts in summer they move into the areas that are frozen in winter, but they return to their warmer winter habitats when the sea ice starts to form again.

Like other Arctic animals, narwhals are very vulnerable and easily affected by pollution of the seas. They are also easily frightened by human intervention and the noise this can bring – they can hear a ship at a distance of tens of kilometres – and they are inclined to flee, which disrupts their usual migration, causing them to make detours and swim further. This tires the animals, especially the young ones.

Their natural predators are polar bears, which catch them in winter when they surface at ice holes to breathe, and pods of orca which work together to

corner and catch a narwhal. Occasionally, a shark or a walrus will attack, but they usually target only the young or infirm narwhals.

Yet another challenge to survival is avoiding being caught under the sea ice and so being deprived of air, causing drowning. It is disastrous when it happens to very large pods of whales, and can occur quite suddenly if, for some climatic reason, the ice freezes over very fast, as has tended to be the case in recent years of global warming.

Recently, a theory has been put forward that the whales become trapped when they are disorientated by acoustic disturbance (for example, nearby seismic surveying). This is considered to be more likely than not noticing the ice freezing, as they would sense it happening and move away.

Narwhals produce calves only every three years or so, but they live to fifty years or more. Thus females are able to produce a number of calves in their lifetime in spite of slow breeding rates. The main threat to their survival is probably the introduction of industry into their habitat, with its accompanying noise and pollution. Another threat is the increasing loss of sea ice which disturbs their migratory habits and leaves them more vulnerable to being caught by their predators.

The population of narwhals is known to be declining, and they are officially considered to be 'Near Threatened', which means that they are at risk of

Fig. 1.41 Narwhal pod swimming in Lancaster Sound, Canada.

becoming endangered if conservation measures are not taken. Officially, the Greenland and Canadian Inuit are allowed to catch a limited number of narwhals according to the IWC's subsistence hunting regulations, but these quota are often ignored. The European Union and the United States have both now banned the import of narwhal tusks, which should mean that the whale is a little more protected. The tusks are very distinctive and of limited use as an ivory material to be worked, so illegal trade is limited to tusks that are to be kept as curiosities.

Tusks

Generally, only the male narwhal carries a single tusk, which erupts from the left-hand side of the upper jaw and spirals anticlockwise. It normally reaches a length of just over 2 m, although tusks of 3 m and more have been recorded.

About one in two hundred male narwhals carry two tusks, the left one usually – but not always – longer then the right.

It is estimated that 2.7 per cent of females also carry tusks, but when they do it is common for them to have two. The females' tusks are seldom more than a metre in length.

A narwhal foetus may have four or more 'teeth' in its jaw. The ones in the lower jaw generally fall out after birth, leaving two in the upper jaw. In the case of the males only the left tooth erupts while the other one remains in the jaw. In females both remain in the jaw. The unerupted teeth are not spiralled.

The tusk emerges when the calf is about two years old, and then grows very fast until it reaches its final length. At this stage the tooth material it is still very thin, and the pulp cavity reaches almost to the tip of the tusk. Over the following years the pulp cavity continues to lay down dentine on the inside of the tusk so that the dentine becomes thicker, but without the tusk becoming wider, until in an old animal the tusk is far more solid, and the pulp cavity has been reduced to a narrow nerve canal.

Fig. 1.42 Section of unworked narwhal tusk, showing spiral morphology. Tusks can reach 3 m in length.

▲ **Fig. 1.43** Unerupted narwhal tusk, which remains in the jaw. Length 18 cm.

In a few cases male narwhals have no tusk. These whales have been noted to be fatter and to have larger testicles, so it is thought that the growth of tusks have something to do with hormonal balance.

It is very common to find male narwhals with broken tusks. Although ivory is a tough material and the narwhal's tusk is capable of bending slightly, its composition, length and shape make it vulnerable to breaking. This can happen in fights with other males, or if knocked too hard against the ice. This would immediately expose the nerve and must be extraordinarily painful.

The purpose of the tusk is still a mystery, although many suggestions have been made. One of the oddest is that the narwhal uses its tusk to spear its fishy prey, before sliding the fish up the tusk and then somehow nibbling it off. However, quite apart from the fact that it would be extremely difficult for the whale to dislodge the fish from the tusk, half of the whales do not have any tusks, yet they do not starve. It is also unlikely that the tusk is for serious sparring purposes, or for trawling the sea bed for food, as both activities would risk breaking it. The most likely purpose therefore, is thought to be that the tusks are for show when mating, and that the whales may also use them in a tactile way for touching each other.

▼ **Fig. 1.44** Cross-section of narwhal tusk showing darker cementum and uneven demarcation line between cementum and dentine. Also visible is the spiralling exterior.

Narwhal tusks are most noted for the anticlockwise spiral along the whole length of the tusk. They are slender: they are seldom more than 6 cm diameter at the base; and they taper gradually to a

Fig. 1.45 Section of narwhal tusk in longitudinal section, showing clear demarcation between cementum and dentine, with irregularities in both owing to the spiral morphology.

smoothly polished, blunt tip. The outside of the tusk is covered in cementum, with indentations following the spiral. There is no enamel.

In cross-section, the tusk is round with an uneven, ridged, outside edge. The cementum has a darker colour than the dentine and a slightly fibrous appearance to the concentric rings. It often displays some cracks, which can reach into the dentine inside. There is a clear and bright demarcation line between cementum and dentine. The dentine appears much finer, often with only a faint indication of concentric rings.

The pulp cavity at the centre continues along the whole length of the tusk. In older animals the cavity gradually closes as more dentine is laid down. This condition can be clearly seen in longitudinal section. Some uneven colouring can also be seen, which vaguely follows the cracks in the cementum, irrespective of the age of the animal.

The unerupted teeth (usually one in the male; two in the female) are straight, slim shafts of dentine, with one bulbous end and tapering to the tip at the other. They are about 16 cm long with no spiral, and they have no pulp cavity. These teeth were made into items such as crochet hooks, and were also hollowed out to make pipes.

See also Appendix I, which lists the main identifying features of the various ivories.

TUSK ANOMALIES

Most tusks are very straight and have a neat, tight spiral, safely enclosing the nerves inside. Occasionally, whales are found with a tusk that has grown in an open spiral or corkscrew which would have exposed the nerve. In other narwhals the tusk is not straight, but has a closed spiral that corkscrews slightly. The reasons for these anomalies are unknown, but it is possible that they may be an hereditary trait as anecdotal evidence suggests that they appear in more than one animal in a pod.

Narwhal ivory is always limited in its usefulness because of its shape. The two anomalies above would make the ivory virtually useless, except as a curiosity.

Uses and Beliefs

Narwhal ivory cannot be carved in the same way as, for example, elephant ivory, because narwhal ivory has a narrow shape and is hollow. Because the spiral of the tusk is decorative, it is usually simply polished and left with little or no extra decoration. It has mostly been used in small plaques set side by side, for example, as a drinking vessel, or left whole and mounted as a curiosity. It was, however, perfect for use as walking canes, where a section would be mounted with a handle of metal or other carved material. Tusks were also left whole and used as the staff of bishops' croziers, and adorned with carved elephant ivory crooks at the top.

Fig. 1.46 Silver tankard covered in narwhal ivory plaques, made in Denmark in 1654 by Jens Jakob Nordmand. Note the spiral morphology.

Narwhal ivory was for several centuries thought to be unicorn horn. The idea of unicorns started about 4,000 years ago, and there were many and varied thoughts as to the shape of the animal and its habitat. As with walruses, early descriptions of narwhals were fanciful as very few people had actually seen one. Thus, narwhals were sometimes described as monsters with huge horns, and some descriptions had the whole length of the animal's body spiralling like the tusk.

The colour of a unicorn's horn was uncertain as well. But by about 1200 the various ideas converged, and it was generally accepted that the horn was somewhere between 1 m and 2 m long, of a pale ivory colour, and that it spiralled. Not only did narwhal tusks fit

Fig. 1.47 Unicorn – the fantasy made real

the description beautifully, but their existence was thought to be proof that unicorns were not a myth but really did exist.

The Norse settlers in Iceland and Greenland soon realized that narwhal tusks could be traded, and began sending them to Scandinavia and other northern European countries, content to allow the recipients to believe them to be unicorn horns. The 'horns' were extremely valuable, and thought to have medicinal and almost magical properties, the most famous of which was their ability to detect poison.

Few people knew where the 'horns' originated and that they might not be from unicorns. Science was in its infancy and people accepted what they were told rather than querying it. Those who realized that the horn came from 'a fish' decided that there must also be a marine species of unicorn.

Further, it was not in the interests of the people trading it, or receiving it as precious gifts, to know the truth. Many cathedrals and royal courts in Europe owned one or more 'horns', and large sums were invested in them. For a while it was believed that they had religious significance, while amongst the nobility

– who were ever-fearful of being poisoned – it became a custom to have a piece of narwhal tusk mounted to stand on the dining table. Pieces were also mounted as pendants to be worn around the neck. The belief was that the 'horn' would start to sweat if it came near poison. Powdered unicorn horn could be purchased in apothecary shops, and from other sources, and there was a lively trade in fake pieces of horn or powdered material which were sold to the unwary.

In Denmark, King Frederik III decided to have a throne made of 'unicorn horn', in a similar style to King Solomon's reputed elephant ivory throne of the tenth century BC. King Frederik was unable to surround his throne with twelve lions cast in gold, and only managed six cast in solid silver before the court's coffers ran dry. Three of those were later melted down to help the cash-strapped economy. To obtain the necessary number of tusks for the throne it is reported that three voyages to Greenland had to be made, and in fact the king died in 1670 before the throne was completed (*see* Fig. 3.15, which shows the Danish throne made with narwhal ivory).

By this time various men of science were querying the existence of unicorns, and also whether the 'horn' had any medicinal properties, although various tests involving poison, powdered horn and cats or scorpions did suggest that it might. A Danish doctor and philosopher, Ole Worm, obtained a whole narwhal cranium complete with tusk, and through his research it was finally decided that the 'unicorn horn' that was being traded and used in northern Europe was, in fact, a whale's tooth. However, he did not completely discount the idea that somewhere, somehow, unicorns might exist in another form.

By the middle of the eighteenth century people had finally accepted that narwhals were not unicorns and their tusks had no medicinal properties, and the fascination with 'unicorn horn' paled. As a result the trade in narwhal ivory almost ceased, and the ivory was left to the local Inuit people, who continued to use it for utilitarian purposes such as harpoon points, handles, or small carvings. Very straight and tightly spiralled tusks could be used as harpoon shafts.

The unicorn is still regarded with fondness, and still appears on various coats of arms including the Royal Coat of Arms of Great Britain, where it is seen in the now more-common shape of a white horse with a spiralling horn emerging from its forehead.

1.7

Sperm Whale

The sperm whale, *Physeter macrocephalus*, sometimes called the cachelot, is the largest of all the Odontoceti, or toothed whales. Males generally grow to a length of up to 16 m, although some have been recorded that were much larger. They normally weigh about 45 tonnes. Females are on average 40 per cent smaller.

Status and Conservation

Sperm whales can be found worldwide in both warm and cold seas, living in groups or 'pods' of either males, or females with young. They live to about seventy-five years of age, and though the females reach maturity at about ten years old, they only calve every five years or so. This gives them an extremely slow reproductive rate.

Little is known about the sperm whale's underwater life. They are, however, magnificent divers. Because they breathe out before diving, the air in their lungs is not affected by altering pressures of diving and surfacing, so they do not suffer from decompression sickness. They can feed at depths of more than 2,000 m, and can stay down for up to ninety minutes before returning to the surface to breathe. This is much further than any human can venture, with or without equipment. Submersibles are now able to dive to greater depths than before, so more research may be forthcoming.

Sperm whales were the favourite catch of the whaling ships from early in the eighteenth century until late in the twentieth, as the head of the sperm whale has a huge cavity filled with 'spermaceti' – a liquid, waxy substance that could be used for oil lamps, candles and lubricants. It was considered to

▲ **Fig. 1.48** Sperm whale skeleton exhibited at the Natural History Museum, London, March 1901. Note the teeth are in the whale's lower jaw.

produce the very finest quality oil and to be superior to mineral oil, which slowly overtook whale oil in popularity. The bones, meat and ambergris (a grey substance that was found in lumps in the abdomen of the whales and has been used as a fixative for perfumes) were useful by-products of the catch, as were the teeth.

Today sperm whales are covered by commercial whaling bans, but still get trapped in deep sea trawler nets. Unable to surface to breathe, they drown. Another hazard for the whale is shipping, with which they can collide as they surface. Further dangers are pollution – they can absorb so much that they become toxic themselves – and ingestion of marine debris.

Years of hunting the whales led to their numbers being very depleted. Today there are less than a third of the numbers that once swam the oceans. They are listed by the IUCN as 'Vulnerable', and as having 'unfavourable conservation status' by other organizations.

Teeth

Sperm whale teeth are the largest among the toothed whales, and can measure up to 20 cm in length. The whale has up to twenty-six teeth on each side of the lower jaw, which fit into sockets in the upper jaw when the mouth is closed. The whale also has teeth in the upper jaw but they are smaller and very seldom erupt through the gums. If they do, they appear as small stumps.

Fig. 1.49 Unworked sperm whale tooth, showing enamel tip, and ring round distal end caused by the wear of the dentine.

Fig. 1.50 Cross-section of a sperm whale tooth taken from the proximal end, showing growth rings in the cementum and darker rings in the dentine, plus the very clear demarcation line between the two.

Sperm whale teeth are slightly triangular in shape in young animals, and tend to be wide at the proximal end. As the animal grows older, the proximal end contracts and becomes much narrower, giving the well-known fat, peg-like shape.

The teeth have a small patch of enamel at the tip and otherwise consist of two types of cream-coloured tooth material, the dentine and a surrounding thick layer of cementum. The two types are separated by a clearly defined interface.

Seen in cross-section, both cementum and dentine display concentric lines, those in the dentine being much clearer. The lines in the dentine can be counted to tell the age of the animal. There is a dark spot in the dentine at the centre of the solid part of the tooth, though nearer to the proximal end it is hollow, owing to the pulp cavity.

Also in longitudinal section, the different structures of the dentine and the cementum are clearly visible, and the cone-in-cone structure in the dentine is also apparent. There is a darker area at the top of each cone in the dentine (which causes the dark line in longitudinal section and the spot in cross-section). The thick cementum shows very little structure. It is also noticeable that the cementum stops short at the distal end of the tooth, which explains the line around the outside of the tooth (which sometimes appears a faint orange-brown colour). This line is one of the aids in the identification of sperm whale ivory. In some older whales the absence of cementum at the tip of the tooth has resulted in the dentine wearing away faster than the cementum, disturbing the natural curved outline of the tooth.

See also Appendix I, which lists the main identifying features of the various ivories.

TEETH ANOMALIES

In some sperm whale teeth the dentine is not a totally smooth, cone-in-cone structure, but includes small globules in the same colour. They may sit loosely in the dentine and can fall away when the material is carved. In carved items which display a lot of the irregular structure, the ivory may possibly be confused with walrus ivory.

It is not known what causes the anomaly. Various suggestions include disease, damage to the teeth, poor diet, and so forth, but it is known that whole pods of whales can share the anomaly, and that affected pods are found in both the northern and southern hemispheres, so it would seem plausible that the appearance is caused by a genetic factor.

▲ **Fig. 1.51** Sperm whale tooth cut in half lengthways, displaying the globular anomaly in the dentine. Note also the clear cone-in-cone structure of the dentine, the dark line through the centre, the thick cementum with growth lines, and the clear demarcation line between dentine and cementum. Length of tooth, 18 cm.

Uses and Beliefs

As already mentioned, the sperm whale was unfortunate to have been born with a square-shaped head which, in an adult, contains nearly two thousand litres of the waxy substance spermaceti that can be rendered to make very high-quality oil.

However, they were no easy prey. Unlike some other whales, sperm whales will fight back when they are caught, and thanks partly to their immense size, have been known to sink a whaling ship or two. It is thought that Herman Melville's famous story of Moby Dick was based on this fact.

Apart from scrimshaw, whale ivory has also been used by the Inuit and Eskimos of the past to make objects for use in hunting (for example, harpoon points). The small figures called *tupilaks* made by the Greenland Inuit are usually carved in sperm whale ivory (*see* Section 3.3 Palaeo-Eskimos, Inuit and Norsemen).

Around the Pacific, sperm whale ivory was used by various islanders to make ceremonial items such as the Hawaiian pendants called *Lei niho palaoa.* On other islands necklaces of whole or split sperm whale teeth were made. The Maoris in New Zealand also made ceremonial pendants. The ivory for all these items was generally collected from beached whales (*see* Section 3.10 Australasia and South-East Asia).

▼ **Fig. 1.52** Inuit carving of a seal, showing the globular anomaly in sperm whale ivory

1.8

Orca

Orca, *Orcinus orca*, are most commonly called 'killer whales', although they are in fact the largest species of dolphin. This becomes obvious when observing their movement as they swim, which has the typical undulating style of dolphins.

Orca males can grow to a length of about 8 m, and weigh 6 tonnes. Females are smaller, at about 7 m and only 4 tonnes. There are many species or sub-species of orca. Some live in cold waters while others live in warm waters; some are carnivorous and eat seals while others only eat fish. All have a striking black and white colouring, although the details in their markings vary from sub-species to sub-species.

Conservation and Status

As with all dolphins, orca are highly sociable animals and live in stable groups. In the wild they live to an age of about fifty years, although some can live a lot longer. In captivity, however, their lifespan is halved. Females mature at about fifteen years and calve every five years.

Mankind is the biggest threat to orcas, through habitat loss, marine pollution, and the resulting diminishing prey stocks such as salmon. A further threat from mankind is the wish to have orcas in marine parks. In the past a whole sub-species was almost wiped out in the Pacific because of over-hunting in one particular area to furnish the demand. Nowadays most of the orcas in marine parks have been captive bred.

Sound pollution is also believed to have a harmful effect on them, and research suggests that sonar from submarines can damage their ears to the extent that a group will beach themselves in order to avoid it.

They were not originally sought by the whaling vessels as they were considered more trouble to catch than they were worth, but as stocks of whales became depleted in the middle years of the twentieth century, orcas became a target as well. Their teeth were a by-product of the catch. Orcas have not been killed purely for their ivory.

Today they are hunted more by 'whale watch' tourism than by people wishing to catch them. However, subsistence hunting of orcas by the peoples of Indonesia and Greenland is permitted.

In some areas orca populations are considered to be sufficiently under threat for them to be put on 'endangered' lists, although in most areas they are listed as 'data deficient'. There is ongoing research into orcas and their behaviour. At present it is not yet known exactly how many species or sub-species of orca there are, nor indeed how to divide them.

Teeth

Orcas have teeth in both the upper and lower jaw, which fit neatly between each other when the mouth closes. The number of teeth varies, but is usually a total of about forty-eight to fifty-two. They are strong and peg-like, and the front teeth tend to slant slightly forward. It has been suggested that this prevents them from breaking when the animal catches large, wriggling prey; however, it could also be a genetic predisposition in some pods. The teeth are not used for chewing, but for holding and tearing the prey.

Seen in cross-section, orca teeth are very similar to sperm whale teeth and display the same fine lines in the dentine and thick layer of surrounding cementum. Enamel may sometimes be present on the outer edge. As with sperm whales, the teeth are oval to round, with few ridges on the outside. The structure is plain to see in longitudinal section, where the cone-in-cone structure of the dentine and the clear demarcation to the thick cementum layer are clear.

Differences lie in the fact that orca teeth are smaller, and the structure is less easily discernible than in sperm whale teeth. On occasion it may even be necessary to dye the ivory to be able to see the structure.

Unlike sperm whale teeth, the cementum on orca teeth reaches the tip of the tooth, so there is no visible ring on the outside where the cementum stops (as there is on sperm whale teeth), and the dentine does not get worn down and the tooth blunted as a result.

Orca teeth have been used as ivory, though not as much as sperm whale teeth because orcas were not targeted by the whaling fleets. There is also more enamel on an orca's teeth than on those of sperm whales, making them more difficult to work.

See also Appendix I, which lists the main identifying features of the various ivories.

Uses and Beliefs

In spite of the name 'killer whale', orcas will very seldom attack people. Their reputation comes more from the way they catch seals, playing with them before killing them, which may seem cruel but is normal behaviour among some animals (for example, cats with mice). They are very clever hunters and a pod will work together as a team, surrounding their prey or driving it up onto the shore. There seems to be no record of a human ever having been killed by an orca in the wild. However, there have been a few attacks by orcas kept in captivity – for example, in marine parks – and some have been deadly. Rather than prove that the animals are dangerous, and taking into account their apparently unblemished record in the wild, it would seem to be further proof that they do not thrive when kept in captivity.

Although, as mentioned earlier, Orcas were not targeted for their teeth, as a by-product their teeth have been used in the same ways as sperm whale teeth; one example is for scrimshaw.

1.9

The Pig Families

The pig family, Suidae, consists of many different species worldwide, including domestic pigs. The best-known from an ivory point of view are warthogs, wild boar, and barbirusa.

On average, suids have about forty-four teeth in total, but ivory comes only from the canine teeth of certain species of the family. In these the canine teeth develop as tusks, which protrude from the mouth, growing in an upward direction. The tusks vary in size, and can also vary slightly in curvature or position in the mouth.

Fig. 1.53
Phacochoerus aethiopicus, desert warthog

Wild Boar

► **Fig. 1.54** Boar's tusk pendant. Note the worn-down tip on the inner side of the circle.

Wild boar, *Sus scrofa*, are found in Europe, around the Mediterranean including North Africa, and in much of Asia. In America they are sometimes called 'razor-backs' and were first introduced there in the sixteenth century.

Wild boar vary greatly in size, depending on where in the world they are found. An adult male may measure from 55 cm at the shoulder and weigh 50 kg, though they can also be twice that size.

They are prey to various animals; in Asia they are a favourite food of tigers. But they are fierce creatures when under threat. Because of their size, the piglets are prey to many creatures, including snakes.

Wild boar are not under any threat of extinction, and are classed by the IUCN in the category entitled 'Least Concern'.

TUSKS

Only the males carry protruding tusks, and these, too, can vary greatly in size. All the tusks are canines and grow out from the face curving upwards, and all four tusks can be worn to sharp points. In the wild boar, the lower tusks are the larger. They are mostly hollow, have enamel on two surfaces, and are triangular to oval in cross-section. The dentine has faint, irregular concentric lines.

The upper canines are smaller. They also have some enamel on the outer surface, are largely hollow, and vaguely oval in cross-section. The dentine has the same faint concentric lines.

See also Appendix I, which lists the main identifying features of the various ivories.

USES AND BELIEFS

Wild boar have been hunted for their meat or as sport, rather than for their tusks, which are a by-product of the kill. Tusks have been used to make jewellery or ceremonial attire, especially in the Pacific Islands where they have been used to adorn breastplates and headdresses.

As jewellery, the shape and relatively thin dentine has limited the possible use of the tusks. In most cases they were left whole and used as pendants, or many were threaded together with the curve inwards to make chunky bracelets.

In Japan, boars' tusks have been made into *netsuke*, with the theme of the piece being dictated by the shape and curve of the tusk. Tusks have, on occasion, been etched and decorated in the scrimshaw style, purely for decoration.

If a boar loses an upper canine, the lower one will continue to grow unhindered until it reaches a full circle. This makes eating very difficult. As the circular tusks are valuable, some captive boar have the upper ones pulled out and are fed a soft diet in order to encourage the circular growth. This process needs a lot of patience as the tusk can take up to eight years to reach the full circle.

Complete boars' heads have been popular as trophies, hung on the wall. Their image has been used in heraldry and appears on many coats of arms.

Wild boar breed very well in captivity and are bred for their meat. There are no restrictions regarding purchase of the tusks and their movement within a continent; however, importing or exporting tusks (for example, from Europe to America, or vice versa) requires appropriate documentation.

Fig. 1.55 Wild boar tusk, allowed to grow full circle.

Warthog

Warthogs, *Phacochoerus africanus*, are native to sub-Saharan Africa. They have some natural predators but, like the wild boar, it is the piglets that are most vulnerable. Also like the wild boar they have been hunted, but again, not for their ivory.

The animals vary a lot in size, and can measure from 65 cm to 85 cm at the shoulder. A fully grown male can weigh anything between 60 kg and 150 kg. The females are smaller.

Warthogs are on the IUCN list as of 'Least Concern' as populations remain healthy.

TUSKS

Warthogs have two pairs of tusks, both of which are canines. All grow out from the face and upwards. Those in the upper jaw are much larger and more curved than the ones in the lower jaw, which are constantly ground down against the upper tusks as the animal eats, so are kept short and at the same time sharpened to points.

Fig. 1.56 Unworked wild boar tusk.

▲ **Fig. 1.57** Cross-section of warthog tusk, showing uneven central section of dentine.

Warthog tusks are bigger than those of wild boar, and can grow to quite a large size. This can cause confusion between warthog tusks and the canine teeth from a small hippo. Their appearance in cross-section does not always follow a definite pattern, but can vary slightly.

The upper canines have an enamel cap which wears away with age. They are slightly dumbell-shaped in cross-section – a rounded-off rectangle with furrows on three sides. They can also be sub-triangular. The dentine displays faint, irregularly spaced concentric lines, and usually a TIZ (tusk interstitial zone) that appears as a dark line and is a result of the dentine becoming more solid and converging as the animal grows, filling the pulp cavity. The tusk can also have an uneven central area of dentine, or a very small TIZ, or it can have a hollow central area.

The lower canines may have ridged enamel on two sides and thin cementum on the third. They may be hollow for much of their length, and in cross-section are triangular in shape. The dentine displays faint, irregularly spaced concentric lines.

See also Appendix I, which lists the main identifying features of the various ivories.

USES

The upper canines of warthogs are larger than those of wild boar, and are more solid. However, their shape is similar and thus limits their usefulness as ivory. The tusks have been made into jewellery or used in ceremonial attire. They have also been carved in low relief in some parts of Africa, mostly for sale to tourists (*see* Fig. 3.25, carved warthog tusk).

The lower canines have been similarly used but are smaller and more hollow, limiting their versatility even more.

The Giant Forest Hog and the Bushpig

There are several species of suid living in Africa, and it is likely that they have all on occasion been referred to generically as 'warthogs'. Among them can be mentioned the giant forest hog, *Hylochoerus meinertzhageni*, which is the largest of the wild pigs, but with slightly smaller tusks than the warthog, and the smaller bushpig, *Potamochoerus larvatus*, which likewise has smaller tusks.

Their tusks could be used in the same ways as the other pig-family members, though their uses would be slightly more limited due to their smaller size.

They are all on the IUCN list as of 'Least Concern'.

Barbirusa

The barbirusa, *Babyrousa babyrussa*, is a cousin to wild boar and warthogs, but is found only in the much more confined area of four Indonesian islands. There are four sub-species, in one of which the males carry the famous and spectacular curling tusks that can grow so long that they almost complete a full circle.

Barbirusa have natural predators, as do other members of the pig family, but as always it is the piglets that are most vulnerable and likely to be killed as prey. Unlike the warthog and wild boar, barbirusa have been hunted almost to extinction by man, for their meat, their pelts, and for their tusks which have been used locally in the making of ceremonial attire, or as gifts, as they have been attributed almost magical properties. Further, commercial logging has reduced the animals' habitat.

Barbirusa are classified by the IUCN as 'Vulnerable' or 'Endangered', and they are protected in Indonesia, but illegal hunting continues.

TUSKS

As with their cousins, barbirusa have two pairs of tusks. The lower tusks are smaller, and curve upwards alongside the head of the animal. The upper tusks appear to grow straight up from behind the nose, and curve backwards towards the forehead. In most animals they curl sufficiently not to touch the head, but in a few unfortunate individuals they can penetrate the skull between the eyes.

There is no enamel on barbirusa tusks, but they have a thin covering of cementum which is longitudinally ridged. The tips are smooth as they do not get worn down.

In cross-section the tusks are almost circular, and the dentine displays the typical irregular, faint, concentric lines of the suids. The tusks are hollow for much of their length, but where the dentine has filled the cavity there is a dark line at the centre, which is the last vestige of the pulp cavity.

See also Appendix I, which lists the main identifying features of the various ivories.

USES

Images of barbirusa appeared on ceremonial masks in Indonesia, where they were intended to instil awe. The tusks have also been made into jewellery or used for ceremonial attire.

1.10

Other Teeth

There are various teeth that, although made of enamel, cementum and dentine, are never worked, and therefore should more correctly be called 'teeth' than 'ivory'. They have been used as jewellery or for their talismanic properties, often as necklaces, but also as part of headdresses, bracelets, shamans' paraphernalia and other ceremonial attire, or even as currency (*see* Fig. 3.41, showing dogs' teeth necklaces, used as currency in Papua New Guinea).

All teeth can be used in this way, and in museums can be found examples of necklaces made from anything from dogs' teeth to bats' teeth. The following are some of the more notable ones encountered.

Red Deer and Elk

Cervids are found in many parts of the world. The European red deer, *Cervus elaphus*, and the North American elk, *Cervus elaphus canadensis*, were formerly thought to be sub-species of the same species of cervid, but recent research suggests that they are, in fact, separate species. They are both listed by IUCN in the category: 'Least Concern'.

The American animal is slightly larger than the European one, but with both it is the two upper canines that are used to make jewellery such as small pendants, or set as cuff-links. The dentition of antler-bearing animals does not usually include canine teeth, but these two animals are exceptions.

Unworked teeth measure about 2.5 cm long, and have a knobbly, rounded appearance. Occasionally the teeth are carved in low relief, but more frequently part of the proximal end is removed to give a better shape and they are left unadorned. When made into jewellery the teeth are mounted in order to hide any vestige of the pulp cavity. There may be little enamel left, although some is often present around the crown. The colour is creamy-white to pale brown, sometimes slightly mottled.

Although of limited use, the teeth are so popular in some areas that imitations have been made in plastic or porcelain. Once mounted, it can be difficult to recognize the fakes, especially porcelain, which has a similar lustre.

Elk teeth were used by the Native Americans as adornment, both as jewellery and also sewn onto clothes.

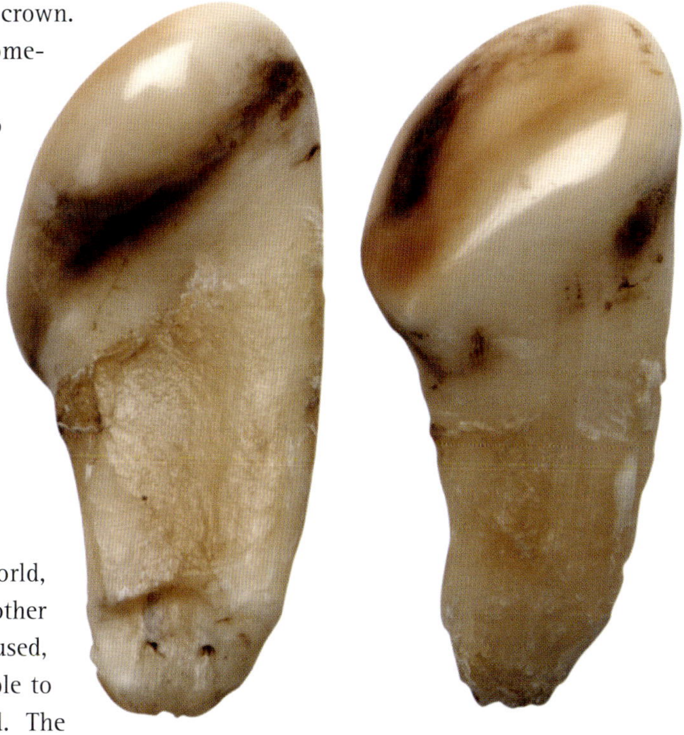

Fig. 1.58 Pair of red deer (American 'elk') teeth in unworked state. Length 1.5 cm.

Bears

There are eight species of bear around the world, ranging from polar bears to pandas. As with other animals, the teeth from any of them may be used, and without DNA testing it may not be possible to see from which species a tooth is derived. The following two examples give an indication of their uses.

POLAR BEAR

Polar bears, *Ursus maritimus*, are called 'Nanuq' by the Inuit. Adult male polar bears weigh on average 455 kg, though they can be much heavier. Females are smaller. They are the world's largest land predators, and an adult polar bear's only enemies are humans, who hunt them and destroy their habitat. The melting ice-cap is also a big threat to the polar bear's survival, both with regard to the individual bears today, and for the long-term prospects of the species. Their average lifespan is less than twenty years.

Polar bears have forty-two teeth in total, made up of molars, incisors and canines, which reflect their carnivorous diet. The canine teeth, which are the most used, are a slightly hooked peg shape, sometimes with longitudinal cracks.

The bears are listed as 'Vulnerable' by IUCN. Today polar bears can be hunted, but only by the indigenous people as part of their subsistence harvest, and only under the very strict regulations of their respective countries. Their parts – fur, teeth and even meat – can subsequently be sold. However, these regulations are under constant scrutiny, and a complete ban will probably be brought in before long.

► **Fig. 1.59** Polar bear tooth pendant. Length 10 cm.

The Inuit have hunted polar bears for thousands of years. The bears were important for the Inuit's survival, and they ate the meat, used the fur for clothing or rugs, and used the teeth for a variety of things such as babies' rattles. They were also hung as talismans on belts, along with reindeer teeth and various other items.

Polar bear teeth have seldom been carved by the Inuit, only polished. This may be because the polar bear has always been regarded with the greatest respect, and was considered to be wise, powerful, and almost human.

BROWN BEAR

The brown bear, *Ursus arctos*, has several sub-species in various parts of the world. It is the bear of childhood stories, and the one upon which the cuddly 'teddy bear' toy has been based. Its slightly dished face adds to the 'cute' look, though in the wild the animal is far from cuddly and can be very dangerous.

An adult so-called 'grizzly' can weigh up to 680 kg, and measure 2.8 m from the tip of its nose to its tail. It can be as much as 1.5 m at the shoulder when standing on all fours. It is listed by IUCN as 'Least Concern'. The main threats to the brown bear are sports hunting, and loss of habitat.

The dentition is similar to that of a polar bear, and is made up of molars, incisors and canines. The incisors are large and strong, but as usual the canines are the largest teeth. They are also slightly hooked.

The teeth and the claws of brown bears have been used as adornment signifying status: the more teeth or claws worn in a necklace, the fiercer the hunter who was wearing them.

Dugongs

Dugongs, *Dugong dugon*, are marine animals living in shallow waters in coastal areas in parts of the Indian Ocean and the Western Pacific.

They are spindle-shaped mammals, closely related to manatees and distantly related to elephants. Dugongs are not related to walruses, although they bear some resemblance.

Dugongs are listed by IUCN as 'Vulnerable'. They are benign creatures, and are not particularly afraid of humans, who are nonetheless their greatest enemy. Dugongs are frequently injured or even killed by shipping and pleasure boats, even in the areas where hunting them is now banned. They are also susceptible to changes in the climate, as cyclones can irreparably damage their feeding grounds. Water pollution is another problem.

Only the males' tusks erupt. They are small, grow from the upper jaw, and in cross-section can be seen to have growth rings from which can be determined the dugong's age. The tusks can be smooth or ridged.

The bones of the dugong are very dense and almost totally lacking in marrow cavities, so they have been worked as much as the tusks. Both have been used for adornment (for example, as bracelets or pendants).

In some countries the tusks have a deep symbolism which affects the way they are utilized, and there are many myths and local legends concerning dugongs. In some areas they were believed to be mermaids.

Seals

Seals belong to the family of semi-aquatic pinnipeds that includes walruses and sea lions.

There are many different species of seals, mostly living in colder waters, with habitats ranging from the Arctic to the Antarctic. Their conservation status varies according to the species.

As the animals vary in size, so too do their teeth, which also vary slightly according to their diet. They can have up to forty-two teeth in total. Common to all seals is a total of four large canines which are peg-shaped to conical, and hollow for much of their length. They are usually used whole and mounted (for example, as pendants). They are occasionally carved in low relief.

Big Cats

The term 'big cat' refers to the large animals of the family Felidae, which includes lions, tigers, cheetahs, leopards, cougars, and jaguars. Some of these

▶ **Fig. 1.60** Crocodile in Northern Territories, Australia.

are listed as 'Endangered' while others are listed as 'Vulnerable'.

Big cat teeth are not often used as jewellery, and they are too small to carve. Also, the parts of a big cat that are mainly used as adornment are the claws. Nonetheless, the four large canine teeth are occasionally seen as adornment signifying status, or for shamans' paraphernalia. The teeth are large, narrow, curved, and very pointed.

▼ **Fig. 1.61** Crocodile teeth, showing large pulp cavity. Average length 1.5 cm.

Crocodiles

Although crocodiles are reptiles, not mammals, the teeth of these creatures have also been used as adornment, in the past as status symbols and today as tourist souvenirs.

Adult crocodiles have over sixty teeth in their jaws at any given time. The teeth are constantly being shed and replaced by new ones, and the process continues throughout life.

As crocodiles are so dangerous, it may be surprising to learn that their teeth have a very large pulp cavity so are hollow for most of their length, regardless of their size. The teeth consist of a hard outer material which is made up of dentine covered with a very thin layer of enamel. Before a tooth is shed, the replacement tooth is already growing within the pulp cavity.

There are many different species of crocodile worldwide, and most of them are protected and listed on CITES Appendix I or II. This status changes sometimes when there are so many crocodiles that they become a danger to humans, and culling then takes place. The crocodile teeth sold as tourist souvenirs are a by-product from animals which are farmed for their meat and skins.

Wallaby and Kangaroo

In Australia, ivory was not used by the native Aboriginals. They were not seafaring people so had little or no contact with toothed whales, and there were no indigenous animals whose teeth were large enough to work or carve.

The native aboriginals used wallaby and kangaroo teeth as adornment. They were used whole, and usually embedded in a larger piece of jewellery, sometimes made from a mixture of resinous substances, which could also be adorned with coloured seeds. Items made included pendants and headbands.

Fig. 1.62 Agile wallaby. Northern Territories, Australia.

Human Teeth

In some cultures it has been usual to use human teeth as adornment – for example, in Polynesia and other Pacific Islands. The custom was also observed by the New Zealand Maoris, and by some African cultures which believed that wearing the teeth of an ancestor imparted the wearer with the wisdom of the departed.

Human teeth are small and covered in a thick layer of enamel. They are therefore not suitable for carving as there would be virtually nothing left after the enamel was removed, as is normal when working ivory.

Human teeth were left whole and strung as necklaces or bracelets, and sometimes dyed. They have also been found embedded in carved objects (for example, Fijian clubs). In Africa whole human jawbones complete with teeth were sometimes attached to ivory hunting horns, as a status symbol signifying victory over an enemy.

Section 2

How and What?

The composition of ivory, its substitutes and fakes
Testing for ivory
Methods used in working ivory and its care

2.1

What is Ivory?

To most people the word 'ivory' means the material that comes from elephant tusks, and it can come as a surprise to many that ivory originates from a variety of other animals as well. Some purists refuse to refer to other types as 'ivory' and insist that true ivory can only be elephant tusk material. While this view is now considered a little old-fashioned, it is true that elephant tusk material is by far the most common form of ivory and, unless otherwise specified, it is to this that the word 'ivory' in common parlance usually refers.

Different species of animals have teeth and tusks of very varied shapes and sizes. They have evolved over millions of years and in most cases reflect the needs of the animal to cope with its habitat and diet. In a few instances the reason for the shape and size remains a mystery, but research continues into the reasons why both the shape and the internal structure vary from species to species.

Some of the first published research into human and animal teeth was carried out by Bernhard Schreger in Germany. He was especially interested in the characteristic pattern of intersecting arcs in proboscidean ivory, and published a paper on the subject in 1800. This resulted in the pattern being called 'Schreger Lines'.

The intersecting arc pattern of proboscidean ivory has had many names since then, including 'engine-turning pattern', 'decussating lines', 'lozenge shapes', and 'curvilinear pattern', all of which are acceptable. The pattern was also for some years incorrectly termed 'Retzius Lines', which was probably a mistake made once and then often repeated. Retzius Lines are incremental growth lines which occur in some tooth enamel, not in dentine. Today 'Schreger Lines' is the most commonly used term, and other descriptions – such as 'intersecting arcs' – are used to ensure that there is no mistake about to which lines reference is being made. 'Intersecting arcs' is the term used in this book.

Owen Lines are similarly named after a nineteenth-century researcher, Richard Owen, who first described the concentric growth lines which are faintly visible in all proboscidean tusks when viewed in cross-section.

The creamy-white dentine which makes up the bulk of tusks or teeth is usually surrounded by a layer of material of very similar colour but of different structure, called cementum. The thickness of the layer depends upon the species. On some teeth the cementum is covered, or partially covered, by a layer of enamel which is an extremely hard tissue, usually slightly paler than the dentine and the cementum, and semi-translucent. In some animals the enamel is worn away during their lives, whilst in others there is very little there to begin with. Any enamel present is always removed from ivory before working the material (with the exception of proboscidean molars) as it is too hard to carve and would also spoil the even colour and polish of the finished item.

Tusks are usually referred to as being 'hollow'. This, of course, is not the case in a live animal, as there is a mass of tissue called pulp inside every tooth or tusk, the size of which depends upon the species. The pulp carries nutrients and lays down the material from which teeth are made, layer upon layer from the inside. In the case of, for example, elephant tusks, they grow longer and wider with age. The cone-in-cone pattern formed by the growth can be seen in longitudinal section of a tusk. Elephant tusks continue to grow throughout the life of the animal, but most teeth and tusks reach an optimal size and virtually cease growing in size or length, although they may continue to produce dentine which gradually fills the pulp cavity inside, making the tooth or tusk more solid.

The converging dentinal layers meet and form different patterns at the centre of the tooth or tusk, depending upon its overall shape and on the shape of the pulp cavity. Thus, for example, in an elephant tusk, which is roughly round in cross-section, there is just a small, dark dot at the centre, while in an animal such as a hippo (whose canine teeth are somewhat triangular or kidney-shaped in cross-section) the dentine converges forming a curved, dark line.

The growth rate of teeth and tusks is recorded in incremental layers in the dentine and the enamel. These layers can be viewed under very high magnification, and are permanent.

Each species of ivory-bearing animal produces teeth or tusks in which the overall shape, size, amount of cementum or enamel, or the internal structure of the dentine and so forth, are characteristic of that animal. Each of these features is useful when identifying ivory, as there can be occasions when the most common identifying characteristics have been covered by marks from the carver's tools or by pigmentation. (For example, see Fig. 1.32 in which the secondary dentine structure is covered by incised lines and pigmentation, and the carving is identified as walrus ivory by its shape.)

Dentine

Dentine makes up the bulk of most teeth or tusks. It is produced in the pulp by the outermost cells called 'odontoblasts'. By weight it is made up of approximately:

- 70 per cent hydroxyapatite, which is a crystalline form of calcium phosphate – chemical formula $Ca_5(PO_4)_3(OH)$.
- 20 per cent organic matter (mostly the protein, collagen).
- 10 per cent water.

The layer of dentine nearest to the pulp is not mineralized and is lacking in hydroxyapatite, but this layer is extremely thin and does not affect the material from the carver's point of view.

Ivory identification depends largely on observing the structure of the dentine, as it varies from one species to another. The visible structure of dentine is caused by the size, orientation, and density of microscopically small 'dentinal tubules', which can measure from about 2 µm to 5 µm across. The size varies with the species. (One µm or micron equals 0.001 mm.) The tubules are embedded in a granular matrix of particles. The density of tubules in the matrix also depends on the species.

Dentinal tubules have been likened to the steel rods in reinforced concrete, strengthening the surrounding material. When a tooth – or worked ivory – is stained, it is the dentinal tubules that absorb the colour.

Near the outside edge of the tusk the structure pattern can be compressed by the new dentine growing from within, so it may become a little flattened. This is clearly visible in cross-section in proboscidean ivory, where the angles of the intersecting arcs are wider at the edge of the tusk than at the centre.

The presence of collagen in dentine gives the material slight flexibility. (Collagen is a protein which is very abundant in the connective tissue of mammals.) That is not to say that a tusk can bend – ivory is a rigid material – but the organic content makes it tougher and less brittle, and so less prone to breaking.

The cavity containing the pulp is usually conical and – in the case of an elephant tusk – reaches about one-third of a tusk. The size and extent of the pulp cavity varies from animal to animal. In some tusks – for example, the narwhal – the cavity reaches most of the length of the tusk. The formation method of the dentine gives rise to the most common pattern of fine, concentric lines seen in ivory in cross-section, or as a cone-in-cone structure when seen in longitudinal section. If damaged, dentine can be repaired by the pulp, though any characteristic structure (for example, the intersecting arcs in elephant ivory) may not reappear.

Cementum

The main purpose of cementum is to hold the tooth in the jawbone, although there is usually a thin coating of cementum on the whole tooth. It is produced in the root of the tooth, inside the gum, by cells called 'cementoblasts' and covers the whole tooth as it emerges from the gum. By weight it is made up of approximately:

- 45 per cent hydroxyapatite.
- 33 per cent organic matter (mostly collagen).
- 22 per cent water.

Although very similar to dentine in chemical composition and colour, it contains more organic matter and is therefore initially softer. However, when it coats the outside of a tooth it can dry out and become harder than the dentine within.

Cementum has a different structure from dentine as it does not contain dentinal tubules. Seen in cross-section it usually occurs in regular concentric layers. It is therefore seldom possible to identify ivory from the cementum alone, although the presence of cementum, its thickness, and the visibility of the demarcation line between the dentine and cementum can give an indication of which species grew the tusk or tooth.

To a limited degree, cementum is capable of being repaired if damaged.

Enamel

Enamel is an extremely hard tissue. Its purpose is to protect the less durable body of the tooth. It has a far higher inorganic content than dentine and cementum, and is roughly composed of:

- 96 per cent hydroxyapatite.
- 4 per cent proteins and water.

While the high mineral content of enamel makes it much harder than dentine, the lack of collagen makes it more brittle and thus more prone to cracking or breaking if knocked. Thus animals with teeth containing very little dentine but a large amount of enamel, such as crocodiles, have a system whereby when they break a tooth, a new one grows to replace it.

When damaged, enamel cannot be replaced.

Pulp

The pulp is a soft tissue lying in a cavity within the tooth or tusk. Its primary function is to form dentine, but other functions include providing nutrients and moisture to the surrounding organic components of the tooth. It is rich in blood vessels, nerves and lymphatics.

The outer layer of the pulp contains the odontoblast cells which produce dentine. When repairing damaged dentine the pulp produces a globular material devoid of dentinal tubules, but it is sufficient for the repair. The irregular material was called 'pith' by ivory traders and carvers.

Hardness

Average measurements on Mohs' scale for the hardness of ivory vary from 2¼ to 3. However, it is difficult to give accurate figures for the hardness of dentine and cementum, as not only will they vary slightly from species to species, but they also vary according to where on a tusk they are measured. Also, while in theory the cementum should be the softer of the two materials owing to its greater collagen content, its hardness depends upon how exposed it is and, indeed, how old it is.

Hardness of any part of the tusk can also depend upon whether the tusk is from a freshly killed animal or from stockpiled material. In the case of mammoth tusks which have been buried in the permafrost for thousands of years, all the material nearer the outside of the tusk has hardened, including the outer layer of dentine, while the dentine inside remains slightly softer.

All dentine nearer to the pulp cavity – especially in a fresh tusk – is softer than that at the outside of the tusk near the cementum. If a thin slice is taken along the length of the tusk and viewed by transmitted light, it can be seen that the area along the centre is less dense and is almost translucent. This area is where the dentine has converged, closing the pulp cavity. When working with the material, a carver will probably try to avoid the area nearest the pulp cavity anyhow as it can also have a slight colour variation. It will only be used when needing a slice of ivory of optimal size, otherwise it is usually hidden at the back or on the inside of a carving. The slight relative density sometimes becomes apparent later, when, for example, the tusk is made into a box which is used for storing coloured materials such as cosmetics, and the slightly softer ivory is more readily stained by the contents (*see* Fig. 1.16).

Enamel is much harder than dentine and cementum, being at about '6' on Mohs' scale. This makes it more brittle and less elastic. Some animals have very little enamel on their teeth when they are born, while others have more enamel than dentine. In adult animals some teeth and tusks have virtually no enamel remaining; others have just a small amount on one side or at the tip, the rest having been worn away.

Enamel is always removed before carving, so the teeth that have a thick covering of enamel are seldom, if ever, worked, and are therefore not usually regarded as true ivory, although they may be used for adornment. Examples of such teeth are human, dog and crocodile teeth.

Toughness

Ivory is relatively soft, but it is not brittle and it is extremely tough. This is why it can be fashioned into such fine carvings (for example, the rigging on a model of a sailing ship).

Another example of how tough ivory can be was shown in 2013, when, in the fight against the illegal harvesting of elephant ivory, a large hoard of confiscated African elephant tusks was burned. In order to enable them to burn more quickly, the tusks were laid on the ground and a large road-roller driven over them. However, the tusks did not break, but remained whole.

2.2

Ivory Substitutes, Fakes and Simulants

Natural Simulants or Substitutes

Ivory has always been an exceptionally highly prized material, and for that reason it has been faked, imitated, and other materials have been used as less expensive substitutes. The use of these materials was not always intended to deceive – for example, many mediaeval carvings were made in bone and were works of art in their own right. In some countries where ivory was not available it became the custom to carve bone.

Artificial substitutes for ivory have been sought for many years, and date back much further than the first plastics. An early example at an attempt to copy ivory used potatoes treated with sulphuric acid.

ARTIFICIALLY AGED IVORY

Ivory can itself be used deceptively, as it can be artificially 'aged'. Today it is done so in order to circumvent the trade bans, but in times past it was done to make an object look antique and thus more valuable.

There are various methods whereby ivory can be made to look old, all of them involving altering the surface patina. The most common methods are to soak the material in a liquid such as tea or tobacco juice, thus discolouring it. Ivory can also be artificially cracked by heating it and have dirt rubbed into the

Fig. 2.1 Elephant ivory disc that has been stained to make it look older. Note the collection of stain in the cracks.

cracks to give a false impression of age. Holding it in the smoke above a fire can also discolour the surface, although the resulting hue is less convincing.

Artificially aged ivory may not crack along the usual lines of old ivory, so the cracks can seem a little incongruous. However, such cracks are not proof of a fake. Any stain is also liable to collect in and around the cracks, which would not happen when the ivory darkens naturally with age.

BONE

The most common substitute for ivory has always been bone. It has a very similar colour and many similar properties, and it can be used in much the same way as ivory. It is fairly common to see bone items sold as ivory in small shops or at antiques fairs because the vendor cannot tell the difference, but there are also plenty of examples of ivory simulants (that is, imitations), sold with the intention of fraud. However, its versatility as a material is mainly limited by its size and the fact that it is hollow.

▲ **Fig. 2.2** Raw bone, cleaned and ready for working.

▼ **Fig. 2.3** Bone box, showing inserted base and Haversian canals. Height 3 cm.

Bone is the rigid material that makes up the skeleton of vertebrates. It consists mainly of calcium phosphate and the organic material, collagen. Bone is a very strong material yet it is relatively light. Weight for weight it is stronger than steel or concrete.

Most bone is round in cross-section, and is compact but hollow. The hollow cavity is filled with a jelly-like substance called 'marrow', and the bone mass contains a system of tiny tubes called Haversian canals which carry blood vessels and nerves. It is the presence of the Haversian canals that identifies bone, although in intricately carved items or inlay they may be difficult to see. Coarse bone is more easily identified, as is an old piece where dirt may have discoloured the canals, making them more visible. The Haversian canals appear as small, irregular lines in longitudinal section, and as dots in cross-section.

Bone carvings have been found that date back about forty thousand years. It seems that bone, antler (*see below*) and ivory were almost interchangeable for small items such as hairpins in Roman times. In mediaeval times bone was carved in parts of southern Europe when ivory was difficult to obtain, and in the fifteenth century the Embriachi workshops in Italy favoured bone over ivory. They produced exquisite items with bone panels carved in relief, which rivalled the best ivory carvings.

Any bone can be worked, though the most usual are the long bones from the legs of cattle. In some areas camel bone has been used, but it has a coarser structure than cattle bone. Sheep bone also has its uses, but the bones are small in size. In a few countries (for example, Tibet) human bones have been carved. Whale pan bones (from the lower jaw of toothed whales) were the largest bones available and were flat, but they are no longer used as whales are a protected species. Bones from whales were worked by scrimshanders on whaling vessels until the whaling bans came into force.

Bone is carved in the same way as ivory, and using the same tools. Like ivory it can be turned on a lathe, carved, incised, stained, and used as inlay. However, unlike ivory it cannot be softened, and it must be very thoroughly cleaned before use. Today this is done with the help of chemicals and heat. Bone is more rigid than ivory, but it is also tougher and less prone to cracking. As with ivory, bone is a poor heat conductor and is therefore suitable for use as handles on metal tea or coffee pots.

Today the use of bone is widespread, and in many places bone has taken over where ivory left off. An example is the carving of Maori *tiki* (symbolic figures) in New Zealand, where the figures carved as pendants have become popular with tourists and are now made in bone instead of ivory.

There is a large production nowadays of small bone plaques and beads imitating ivory, for use in inexpensive jewellery. The beads are hollow but are capped at both ends so that they appear to have been carved from one solid piece.

Few actual tests will show the difference between ivory and bone. Both materials fluoresce a chalky blue colour if put under a UV light in a dark environment. A test to judge the melting point or the smell emitted by burning a scraping of the material will not differentiate between them, either. Where there is real doubt, specialized laboratory tests such as DNA or ZooMS may be needed (*see* Section 2.4 Testing Ivory – Advanced Testing Methods).

However, the following visual characteristics will, in most cases, be sufficient:

Fig. 2.4 Bone bead. The ends are glued in place.

- Shape and curve. In contrast to ivories, most bone is hollow in cross-section, but it is straight longitudinally, while many ivories are slightly curved.
- Lack of identifying ivory structure, e.g. intersecting arcs, secondary dentine, fine lines, or demarcation line between cementum and dentine.
- The presence of Haversian canals, which appear as irregular, fine lines in longitudinal section, and dots in cross-section.
- Depth of relief of carving. High relief can suggest that the item is not hollow, so it is not bone.
- Quality of carving, especially in more modern pieces. Ivory tends to be of higher quality.
- Possible composite nature of an item, hiding the fact that it is hollow
- Although, under specialized UV light bone fluoresces a chalky blue, as does ivory, reconstituted bone materials have less fluorescence. (Plastics generally do not fluoresce.)

ANTLER

Antler is a very fast-growing form of mammalian bone, but although it might therefore be classified as bone, it is always referred to by its own name. Antlers grow in more or less symmetrical pairs from the bony protuberances or 'pedicles' on the heads of Cervidae, or members of the deer family. They vary in shape and size, and are, with the exception of reindeer and caribou, only carried by the males. Antlers are shed annually and are regrown.

Antler has many of the usual characteristics of bone, although it has a different shape. The centre is not hollow, but is filled with a spongy-looking honeycomb of cancellous bone. The tips of the tines on antlers are round in cross-section, but the rest is only round in certain species. Other species have so-called 'palmate' antlers which have flat areas between the tines, like the palm of a hand with fingers. The outside surface of most bone is smooth, while

Fig. 2.5 Raw antler: (left) a burr, and (right) a slice in cross-section. Both approx. 6.5 cm across.

the outside surface of antler is heavily ridged. For much of the time that the animal carries its antlers they are covered in 'velvet' – a soft, furry skin. Beneath the velvet is a mass of blood vessels, which cause ridges as the antler grows.

When an antler is shed naturally, its base – called the burr – is rough and convex, and cannot be polished. However, it is sometimes used as a decorative item in its own right.

Antler has been used in many of the same ways as other bone, though the ridged surface make it less easy to carve in relief as it first needs to be polished smooth. It has therefore been popular for use as knife handles, where the ridged surface is left intact and gives a good grip.

Nonetheless antler was used in Roman times in the same way as bone and ivory, and in mediaeval times it was popular for use as powder flasks – some of which were beautifully carved in relief – and for inlay on firearms.

It has never been necessary to kill the animal to use antler, as antlers are shed annually. Nevertheless, a lot of antler material comes from farmed animals killed for their meat, or animals killed by sports hunters.

As antler is a form of bone, simple tests will not differentiate between the two. Antler and ivory differ in much the same way as bone and ivory (*see* Bone, *above*). Extra observations regarding antler are as follows:

- Antler is not hollow in cross-section, but has a spongy-looking central core surrounded by compact bone.

Fig. 2.6 Pendant carved in antler. Height 5 cm.

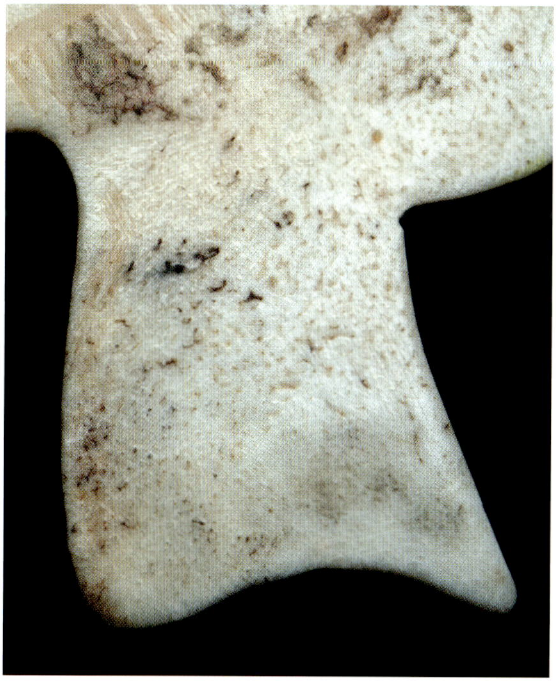

Fig. 2.7 Detail of reverse of antler pendant, showing the rough surface where the central core has been ground away.

- Unless it has been polished away, antler has a ridged outer surface.
- Antler displays Haversian canals.
- The colour of antler varies. It can be a creamy-beige colour, but that from reindeer and caribou is a pale grey.

ODONTOLITE

Odontolite, which is Greek for 'tooth stone' and is known as 'bone turquoise' or 'fossil turquoise', is ivory which is simulating turquoise. Found near Auch in the region of Gers in France, for many years it was thought to be mammoth ivory which had been coloured by natural processes. It was believed that it had been buried in a geological stratum where the earth contained chemicals that had stained it an attractive turquoise blue.

Odontolite is now known to be mammoth ivory from Miocene geological layers that has been artificially heat-treated, altering its colour to a turquoise blue. This was probably done by the Cistercian monks living in the area, who discovered the process in the Middle Ages. They practised agriculture, but were also known to study metallurgy.

Trace amounts of various minerals have been found to be present in the material, although not the vivianite originally thought to be influential in producing the blue colour. Research has not so far been able to identify exactly how and why the treatment worked, although it seems that if mammoth ivory from the area is subjected to 600°C in an oxidizing atmosphere for eight hours, it can turn a brilliant turquoise colour. (Bone is also sometimes found that has been similarly treated.)

In order to carry out further research more samples are needed, as known examples are few and far between. Undoubtedly there are many in private collections that are wrongly identified and believed to be turquoise. For example, odontolite gems sometimes come to light set in Victorian jewellery.

The material is usually cut *en cabochon* and mounted, and careful examination is needed to discern whether or not there is a pattern of intersecting arcs (Schreger Lines) which would identify the material, as these do not occur in the mineral turquoise. Odontolite is also lacking in the dark areas of matrix usually seen in turquoise.

VEGETABLE IVORY

The material called 'vegetable ivory' is so named because it is of plant origin, and is used as an ivory substitute.

It comes from palm nuts, and can be obtained from various species. By far the most common is the tagua palm nut – other common names for it are corozo nut, or ivory palm nut. The nuts are of the genus *Phytelephas*, (meaning

◀ **Fig. 2.8** Tagua nuts. Clockwise from left: with husk; husk partially removed; fully polished; a broken piece. The whole nuts are about 5 cm long.

'plant elephant'), which has several sub-species. The one usually used as vegetable ivory and cultivated for the purpose of harvesting the nuts is *P. macrocarpa*, but *P. aequatorialis* is also used. Taguas are stemless palm trees that grow in the rain forests of Ecuador, Panama and Columbia in South America. There has been limited success in cultivating *P. macrocarpa* in Florida.

The biggest production of vegetable ivory is in South America, though it can also be obtained from the doum palm nut, *Hyphaene thebaica*, which is a palm tree growing in East Africa, or from the Far Eastern ivory nut palm *Metroxylon amicorum*. Both these species produce single-seeded fruits, whereas the tagua palm produces between four and ten seeds in each fruit. The Far Eastern palm is listed as 'Vulnerable' on the IUCN red list.

From pollination to harvesting, tagua nuts take nearly a year to form. The fruit – which is larger than a human head – has a thorny outer casing, inside which is a yellow flesh containing the creamy-white nuts encased in immensely hard shells. Each nut is about the size of a chicken's egg, and is covered in a brown husk. The fruit are eaten by animals, especially deer and rodents.

Tagua and the other, similar nuts are almost pure cellulose, a complex carbohydrate which is the main constituent of plants. When they ripen they are soft and edible, but when dried (either naturally or by artificial means) they become very hard and rigid in texture. In drying they can crack in the centre. This is a hazard that is not

▼ **Fig. 2.9** Doum palm nut.

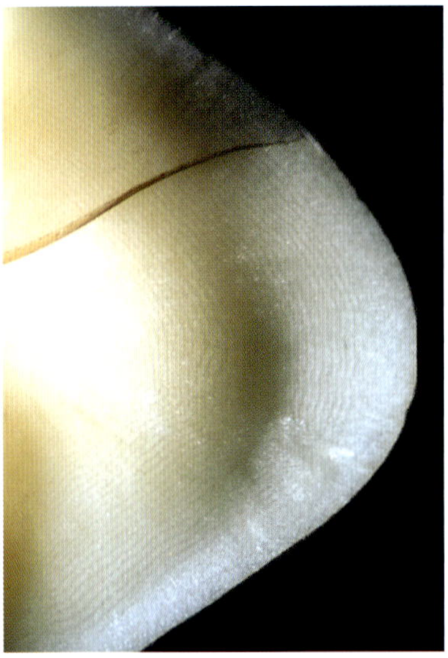

Fig. 2.10 Detail of a tagua nut, showing the cellulose structure.

Fig. 2.11 Tagua nut *netsuke*. Areas have been dyed and the husk has been left on part of the base as a decorative effect.

infrequently encountered when carving the nuts, and only becomes apparent when carving is underway.

When completely dry, the nuts can be intricately carved and polished to a high, waxy lustre. Their natural colouring when new is very close to that of ivory, though they darken to a pale beige colour with time. The nuts can also be dyed or painted, and part of the brown husk is often left on a carving as part of the decoration.

Doum palm nuts have been found in the tombs of Egyptian pharaohs, leading us to believe that they had some sacred purpose. However, it is very unusual to find nuts that survive for such a long time, so the known history of vegetable ivory is concentrated more on the past couple of hundred years.

Tagua nuts first appeared in Europe in the early nineteenth century when they were imported from South America and used in the manufacture of toys, small carvings, and items such as umbrella handles, where they were a good imitation of ivory. The finished object was limited in scope by the size and shape of the nut, but the material could be turned to create a thread, so two or more nuts could be screwed together. Thus items such as sewing needle cases and thimble boxes were popular in Victorian times; indeed, the nuts are so hard that even thimbles could be made from them. Other uses of vegetable ivory included dice and jewellery, all of which were imitating ivory.

However, the greatest use for tagua nuts was in the button industry. It started in the mid-nineteenth century, and factories opened in England, France and America, among other places. With the advent of plastics the button industry declined, and today the nuts are mostly used to make fashion jewellery or small carvings. Tagua nut *netsuke* are produced by the thousand in China, and usually depict animals.

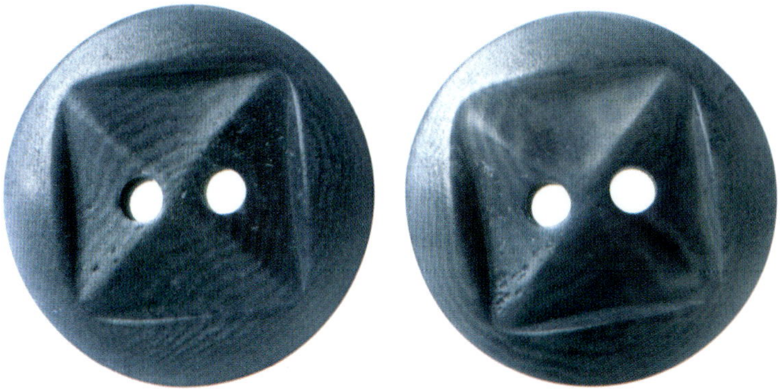

Fig. 2.13 Buttons made from tagua nut. Note the 'moiré silk effect' caused by the structure of the cellulose.

Fig. 2.12 Victorian sewing needle case made from two tagua nuts.

Tagua nuts are proving to be an aid to the conservation of the rainforest, as cultivating them discourages the clearing of these threatened areas. As a substitute for ivory, tagua nuts may also – to a small extent – help to protect ivory-bearing animals such as the elephant.

A test for vegetable ivory is to place a drop of sulphuric acid on the surface. It will react with the cellulose and turn the spot pink. The colour is permanent and cannot be washed or polished off, making it an extremely destructive test which should only be used in exceptional circumstances.

Another test is to judge the fluorescence under UV light. In its natural state vegetable ivory fluoresces a chalky blue – much like ivory – however, it is often dyed or coated which can affect the result, making the test somewhat unreliable.

Vegetable ivory has a distinctive structure which can be seen under magnification. With time the polished surfaces can display a moiré silk effect. Identifying features are:

- Lack of structure typical to an ivory.
- Structure of closely packed, tiny spots of equal size – the plant cells – often seen in wavy bands.
- 'Moiré silk' effect, usually seen on the flat surfaces of older items that have been dyed and polished.
- Possible random cracks at the centre of the nut.
- The creamy-white colour darkened to a woody beige in old carved items.

HORNBILL IVORY

Hornbill ivory is not a true ivory substitute, but is a rare gem material in its own right. It comes from the casque (a large protuberance covering part of the head and bill) of the helmeted hornbill, *Rhinoplax vigil*, which is native to Borneo, Sumatra and in the Malay Peninsular. The term 'ivory' in connection with the hornbill possibly originated in the Far East, where 'ivory' was added to various organic materials of a pale colour that could be carved.

Fig. 2.14
Helmeted hornbill head, showing casque (here carved).

There are many different species and sub-species of hornbill, and several of them have a casque. The helmeted hornbill, which is the largest of the species in Africa, is unique in having a solid casque, with a straight, vertical front. Other hornbills have partially hollow casques that are curved or rounded at the front.

The casque is made of keratin, which is a horny substance very prevalent in the animal kingdom. Keratin is softer to carve than ivory, but the pale, creamy-yellow colour has been regarded as similar to ivory and it takes intricate carving.

The top and sides of the casque are an orange to red colour. This is the result of the bird preening its feathers with the secretion from the preen gland, which dyes the surface of the casque. The front of the casque is not affected; nor, usually, is the bill.

It is uncertain why the bird has the large, solid casque, which can account for as much as 10 per cent of the bird's total weight. However, it is known that the male barricades the female into the hollow of a tree while she sits on the nest, leaving only a small aperture through which to provide her with food. To do this the birds use their casques to tamp down earth and plant matter to block the entrance.

The helmeted hornbill appears on the IUCN red list as 'Near Threatened', and is listed on Appendix I of CITES. It has been hunted for about two thousand years by the local indigenous people for its casque, and for its two very long tail feathers, which have been used in ceremonial headdresses. The carved casques were used as adornment, hung from a belt around the waist, sometimes with tassels attached. Today some hunting continues locally, but the worst

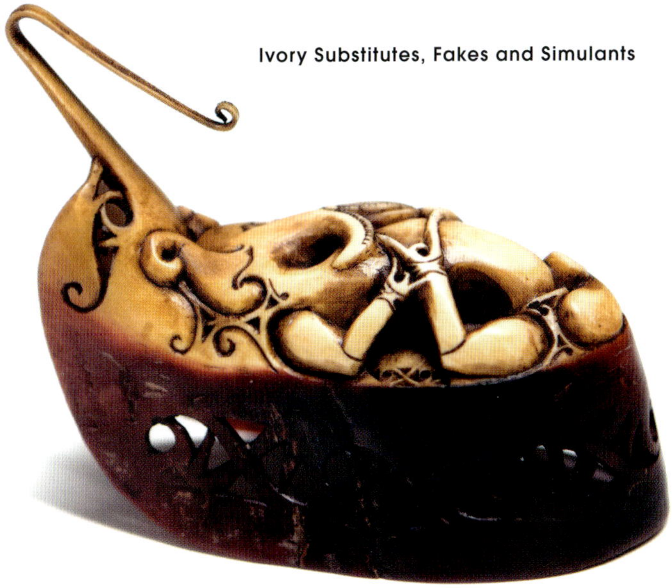

threat to the bird is deforestation, and forest fires which further destroy its habitat. The bird is already extinct in Singapore.

In some cultures the bird was thought to guard the river between life and death, but it was also believed that a ring made from the casque could detect poison by changing colour. In the Malay Peninsula the bird was given the name 'kill your mother-in-law bird' because it could make a cry sounding like maniacal laughter.

Carvings made from hornbill ivory have been found in Borneo which are thought to be almost two thousand years old. The first record of the material being used in China is from the tenth century, although it may have been used there before that time. The Chinese made beads, belt buckles, brooches, snuff bottles, personal seals and suchlike, and sometimes called the material 'golden jade'. The Japanese also admired the material, and used it to make *netsuke*.

Fig. 2.15 Carved hornbill ivory ear pendant. Part of the bill has been used and bent into shape to form the hook.

When carving, part of the red surface is usually incorporated in the pattern, but sometimes it is removed completely, which can make the material difficult to identify. The complete lack of dentinal structure and its slightly golden colour make it unlikely that it will be confused with ivory, and it may be possible to see the laminated, horny structure of the casque, especially on larger surfaces.

Because the casque and bill are largely made of keratinous material, they can be moulded into shape with heat and pressure. Although this is seldom done to the body of the work, the bill is sometimes bent to form a hook so that the casques can be worn as ear ornaments, or suspended from a belt, or similar.

Hornbill ivory is rare, and part of its attraction is the contrasting red surface area which is, whenever possible, incorporated in the worked item. Being rare and sought

Fig. 2.16 Detail of carved casque, showing lamination of keratinous material at the edge of the red area.

after, some 'hornbill ivory' that is offered for sale has been enhanced, altered, or even faked. For example, it is possible to come across bead composites which have had a false red area added to the surface. This may come from a small piece of hornbill casque, but it may also be plastic.

In some cases none of the material is hornbill ivory, and the whole object is an imitation. The most common hornbill ivory simulant is golden plastic with an area of red (also plastic) on one side or surface.

Viewed under magnification some identifying features of hornbill ivory are:

- Laminated, horny structure in the material, best seen by transmitted light and indicating hornbill ivory.
- Blurred line between the red and yellow areas (seen only under high magnification) indicating hornbill ivory.
- Sharp demarcation between the two colours, indicating a doublet, possibly (at least in part) not hornbill ivory.
- Swirls of colour at the demarcation between the colours, indicating a plastic imitation.
- Air bubbles, indicating plastic.

Also, without magnification:

- Under UV light the pale area of the casque fluoresces a pale blue, similar to ivory and bone, while the red area appears much more dull.
- Red colour in more than one area – the red surface is naturally restricted to the surface of the casque and would not, for example, appear on two opposite sides of a small bead.

Fig. 2.17 Plastic box imitating ivory. Note the thin walls and the dull, faded colour.

Synthetic Ivory Substitutes

Plastics are generally much lighter than ivory, and while this did not matter too much with imitation ivory jewellery, it was important to get the weight more correct in carvings or scrimshaw – especially if the intention was to produce a fake and pass it off as ivory. Therefore, a filler of some sort would be added to the plastic. In some cases the result was a material that had the wrong feel because it was too heavy. Old plastic items can lose their original colour and go a dirty, dull cream colour.

Fig. 2.18 Knife handle, showing parallel lines imitating ivory structure.

Objects such as pots are often also moulded with very thin walls. Turned ivory would not be as thin in a utilitarian object.

Most jewellery and 'carvings' were made by moulding the plastic material, but some were made by the extrusion method of forcing the plastic material through a nozzle before it set, into long cylinders or blocks. The advantage of this method was that the plastic could be made up of two types of material, one opaque and one translucent, packed closely together and extruded in layers or a chequerboard pattern. This was used to imitate the structural pattern of elephant ivory, and was particularly popular for cutlery handles.

When the cutlery was examined the weight of the plastic was disguised by the weight of the metal. However, the parallel lines in the plastic imitation were always too precise and regular, and seen in cross-section (for example, at the end of a knife handle), either lines or a chequerboard pattern would be visible, rather than the intersecting arc pattern of elephant ivory.

Fig. 2.19 Imitation ivory and amber beads. Note the lack of intersecting arcs at the end of the bead.

CELLULOID

Celluloid is an early, semi-synthetic plastic, based on the plant material cellulose, usually from wood or cotton, which was modified with nitric acid and plasticized with camphor.

Celluloid was created in the late 1860s and patented under that name in 1870. Other names have included 'Parkesine' and 'Xylonite'. It was the first successful synthetic ivory imitation, and was used for many items from jewellery to billiard balls. It was the most commonly used plastic for imitation ivory as it could be moulded with great detail, resembling carved ivory.

Like many plastics, celluloid can degrade with time and nothing can be done to halt the process. Celluloid ivory imitations eventually start to turn yellow and to warp. Early plastics should never be stored in sealed containers,

▲ **Figs 2.20** & **2.21**
Front and back of a celluloid brooch. Note the fine detail in moulding and the slight distortion and discolouration on the reverse as the material begins to degrade.

such as plastic bags, as this encourages the degrading process. If, however, they start to degrade in such a container, when the container is opened it is noticeable that the item has started to emit a smell of camphor (a smell similar to moth balls). The item will finally simply fall to pieces.

Another big drawback to celluloid was that it was highly combustible. For this reason a similar plastic was evolved, using acetic acid instead of nitric acid. Although it still degraded (though with a smell of vinegar instead of camphor), it was much safer to use as it was not highly combustible.

Note: When testing ivory imitations, the 'hot point test' should be exercised with great caution in case the material proves to be celluloid, which is highly combustible.

CASEIN

Another early, semi-synthetic plastic is casein. The original polymer was invented at the end of the nineteenth century and marketed under names such as 'Galalith' and 'Erinoid'. It was based on the main protein in milk (also called casein), which was reacted with formaldehyde to produce a hard material. It was used a lot in the first half of the twentieth century and is still used today for the manufacture of buttons. Casein is considered the most beautiful of the early plastics because it was very versatile and, by altering the additives, could be made to imitate many different organics, including mother-of-pearl. Because it could be made in pale colours it was suitable as an ivory imitation.

Like ivory, casein is hygroscopic, although constant high absorption of moisture and drying out will eventually cause it to crack. It seldom degrades as completely as celluloid. Because it is hygroscopic, casein is more suitable than other plastics for imitating ivory keys on a piano, where ivory has been difficult to replace partly because most plastics are too slippery. The Yamaha

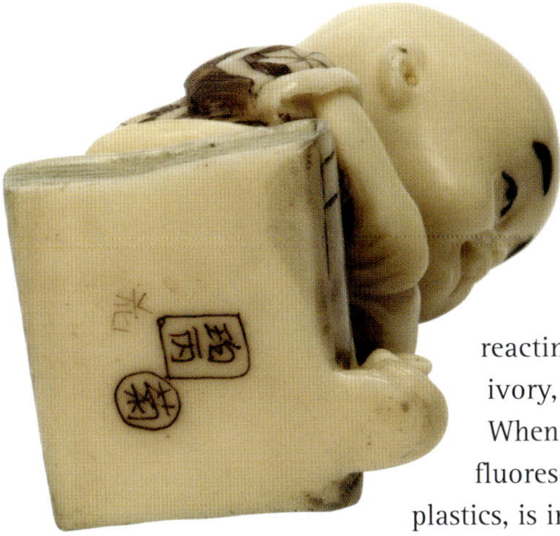

Corporation of Japan patented a form of hardened and stable casein for use on their keyboards, and called it 'Ivorite'. Various other mixtures using casein have also been developed.

Casein is unique among plastics in reacting to UV light in the same way as ivory, and fluorescing chalky blue colour. When mixed with other components, casein fluoresces a darker blue or, similar to other plastics, is inert.

◄ ▲ **Fig. 2.22** & **2.23** Small moulded figure, probably made from epoxy resin. The 'undercutting' and lack of mould marks suggests a soft mould was used.

EPOXY RESIN

There is a wide range of epoxy resins on the market today. As an ivory imitation it has the advantage that it can be moulded in a pliable mould which can be pulled off, rather than a rigid mould that must be removed in parts. Thus it is possible to mould with far greater detail and with some 'undercutting'. It can also be weighted to resemble the weight of ivory. It is very suitable for items such as imitation sperm whale tooth scrimshaw.

ELFORYN

This relatively new plastic was created specifically to be an ivory simulant. It is creamy-white and lacking in any structure, but it can be worked in very

◄ **Fig. 2.24** Old plastic moulding, where the mould was in two halves, leaving marks from the join along the centre of the object.

much the same way as ivory. It can be turned on a lathe and carved, using the same tools as would be used for real ivory. Furthermore, it is said to feel very like ivory in the process, and to be very satisfying to work with.

Elforyn is made from a combination of plastics and minerals, but the exact formula is known only to the manufacturers. It can be delivered from the manufacturers in Germany in many sized and shaped rods or blocks which are made by the extrusion method. It can also be supplied partially formed to minimize waste. It has been used with success in the manufacture of billiard balls.

Elforyn fluoresces a slightly darker blue colour than ivory under LWUV.

POLYESTER

One of the most common and versatile of plastics, polyester can be produced in pale and opaque varieties. It can be used to make individual, hand-made items as well as massed-produced ones, so can imitate hand-carved ivory.

OTHER SIMULANTS

There are many modern plastics on the market today, and new ones are constantly being developed. It is usually sufficient to know that an object is not ivory, and the actual nature of the synthetic resin is seldom questioned.

Now and then an object is encountered that is made of powdered bone mixed with a polymer. This lacks any type of characteristic structure associated with ivory, but it can fluoresce under UV light, confusing the issue.

There are also a few natural materials that might, on occasion, be mistaken for ivory, or at least be considered when identifying an object by process of elimination. Of these should be mentioned shell, which often has very irregular layers of different opacities, and which displays a structure of extremely fine, regular striations if viewed under 10x magnification. It is also cold to the touch as it consists mostly of calcium carbonate.

Also made of calcium carbonate and cold to the touch is coral. It is unlikely that white corals will cause confusion, with the exception of a white *Corallium* coral (also known as precious coral). However, this also has a structure of fine, regular striations, which are visible under 10x magnification. The striations in coral are not as fine as those in shell.

Fig. 2.25
Composite of bone and a polymer, imitating ivory.

Because of the very limited size and shape of shell and coral, there are few occasions when they could come into consideration. The most likely is a carved cameo or other piece of jewellery. Both shell and coral have a vitreous or glassy lustre, whereas even highly polished ivory has a softer look. Both also feel very hard, and lack the tactile qualities of ivory.

IDENTIFYING FEATURES OF PLASTIC SIMULANTS

When examining an object, the following can give some indication that it could be plastic and not ivory, bone or antler.

- Little or no fluorescence under UV light. (*Note:* Casein fluoresces chalky blue in a manner similar to ivory and bone, and reconstituted ivory or bone mixed with a polymer can also fluoresce.)
- Lack of characteristic ivory structures such as intersecting arcs, or the Haversian canals of bone.
- Structure that is too precise, e.g. very regular parallel lines in longitudinal section.
- Warm to the touch, and slippery feel.
- Light in weight.
- Thickness of moulding (for example, in small jars and pots for use on a dressing table).
- Colour faded to dirty, dull cream colour.
- Presence of mould marks or air bubbles.
- Surface lustre too shiny and not 'resinous'.
- Melts when touched with a hot point, giving off an acrid smell, or one of camphor or vinegar. (*Note:* Care should be taken with this test as celluloid is highly combustible.)

See also following sections 2.3 and 2.4 on various tests for ivory.

2.3

Testing Ivory – Simple Testing Methods

The methods and observations given below are the basic tests for ivories, their simulants, and their fakes. Older ivories, especially those from archaeological digs, would in all probability be subjected to more extensive tests.

The usual gemmological tests have little relevance for ivories, as they are created mainly for testing small gemstones. Refractive index (a gemmological test showing the refraction of light through or off the surface of a specimen), is relatively useless in the identification of ivory, and could damage the surface of the item because of the liquids used in testing. Also, it would be virtually impossible to test larger ivory items on a refractometer, which is designed to test gemstones.

Specific gravity is not a good test either, as it cannot be carried out on a large object without immersing it in water. This could cause serious damage to ivory as it is hygroscopic. Knowing the exact specific gravity would be of little use anyhow, though it is sometimes necessary to make a rough judgement of the weight of an item when deciding whether it is true ivory or an imitation.

Hardness is very pertinent to ivory, but more from the carver's point of view than for the purposes of identification. Also, although a hardness figure measured on Mohs' scale is often quoted for ivory, it actually varies slightly from species to species and with the age of the tusk or tooth. It also varies depending on from which part of the tusk or tooth the measurement is taken (*see* Section 2.1 What is Ivory?).

The following, therefore, are the more practical tests for ivory today.

- Visual and touch.
- UV light.
- 'Hot point' test.

Visual and Touch

Without doubt, the most important test for ivory is observation. Examination of an object under a good light is often enough to tell the nature of the material from which it is made. Unlike many other materials, ivories can have a characteristic structure which can be seen with the naked eye. For example, elephant ivory has a structure of intersecting arcs; mammoth ivory has the same but the arcs are narrower; walrus has a bubbly-looking secondary dentine; and so forth (see also the various chapters concerning the animals and their ivory's identifying characteristics.) There can be obvious features that make it clear that an item is not made of ivory. For example, the presence of the Haversian system, indicating bone, and air bubbles or moulding marks, which would indicate a man-made simulant.

The feel of an object can also give an indication of its nature. Its weight in the hand (that is, its heft) can help with its identification. As an example, plastics are lighter than ivories, unless they have been weighted with a filler during manufacture. Plastics are also warmer to the touch than ivory, although ivory feels warmer than a mineral.

The surface of ivory can be polished to varying degrees, so it can display anything from a dull lustre to a high polish. The lustre is usually described as 'resinous', meaning that it has a slightly waxy look rather than a very shiny one.

Ivory that has not been kept in a good environment may become a little dry and dull, and lose its normal smooth, silky feel. When comparing highly polished ivory to plastic, ivory feels silky while plastic feels slippery. This is a consequence of ivory being hygroscopic, while plastics (with the exception of casein products) are not. It is worth remembering, though, that on occasion ivory can be coated with a plastic or lacquer in order to protect it, for example, in pots for cosmetics, in which case the surface feel is misleading.

Problems are likely to arise when dealing with an item that is too small or too intricately carved for any structure to be visible, such as marquetry and other forms of inlay, or a piece of jewellery in a setting.

Sometimes it is a process of elimination that gives the answer. For example, the presence of intersecting arcs would mean that the item cannot be walrus ivory but must be proboscidean: mammoth, mastodon or elephant. The colour should be creamy-white, or – in the case of older ivories – a darkened version. If it is a dull, beige colour resembling wood it could be 'vegetable ivory' and further examination will be necessary under a 10x magnifying lens or a microscope.

There are times when it is not possible to tell whether the material is ivory or bone, or which type of ivory has been used. A small carving may not display any structure at all. The complete absence of any Haversian canals indicate that it is ivory, but it could be from the centre of an elephant tusk where the structure is less apparent, or hippo ivory, or from the edge of a walrus tusk. It is often sufficient simply to know that it *is* ivory and not an imitation. If it is imperative to know which type, more advanced testing can usually be undertaken by a specialized laboratory (*see* Section 2.4 Testing Ivory – Advanced Testing Methods).

Artificially aged ivory can also sometimes be determined by observation. There can be signs of the material having a surface dye or tint, where the colour has collected in cracks or fissures. The cracks themselves may run in the wrong direction and not follow the natural lines of weakness that occur when the ivory dries out.

Ultraviolet Light (UV)

Ultraviolet light (UV) is a radiation in the electro-magnetic spectrum which is invisible to the human eye. The wavelength is shorter than that of visible light, but longer than X-ray. It is sometimes termed 'black light'.

UV testing shows how a given substance reacts to ultraviolet light, and whether it will fluoresce or not. Fluorescence is caused by an item absorbing wavelengths in one part of the electro-magnetic spectrum – in this case UV – and re-emitting them in the visible wavelengths. Gemmological tests use either 'short wave' or 'long wave' UV (SWUV or LWUV) as materials may react differently to them. Ivory is usually tested under LWUV, and it is the collagen content that causes it to fluoresce.

When ivory is illuminated by LWUV in darkened surroundings, it is seen to fluoresce a chalky blue colour. Bone reacts in very much the same way as ivory. Most plastics are inert under UV (that is, they do not re-emit any light) although the plastic called casein – which is made from milk proteins – reacts in much the same way as ivory. Historical repairs to an ivory item that are not noticeable under normal lighting may sometimes show up under UV light.

UV testing will not give a definitive answer regarding the nature of the specimen being tested, but it can give a helpful indication.

Fig. 2.26 Three bangles seen under LWUV light. Left to right: elephant ivory, mammoth ivory, and plastic imitation.

Note: Care should be taken if attempting to use UV light as it is carcinogenic if used indiscriminately. It must never be shone into the eyes or directly onto the skin.

Fig. 2.27 Three figures seen under LWUV light. Left to right: bone, walrus ivory, and plastic imitation.

Hot Point or Burning Tests

Hot point or burning tests demonstrate the melting or burning point of a specimen, and the fumes emitted can indicate its nature.

The test can be carried out either by taking a tiny scraping from an area of the specimen that will not be seen (for example, the drill hole of bead) and holding it over a flame, or by heating a small, sharp point such as a pin until it is white hot, and holding the point against the specimen – again taking care to do so in an area where any ensuing damage will not be seen.

Ivory scrapings will burn when exposed to high temperatures, and a hot point held against an ivory object may leave a small scorch mark. By contrast, plastics will melt, either as a scraping or when the hot point is applied, in the latter case leaving a hole in the specimen as the point melts its way into it. Furthermore, plastics emit an acrid smell when heated to a high temperature, while ivory does not.

This is a destructive test as it damages the specimen being tested. It cannot differentiate between bone and ivory as they react in the same way.

Note: This test is not advised unless undertaken with utmost care, as the early plastic most commonly used to imitate ivory is celluloid or cellulose nitrate, which is highly combustible.

2.4

Testing Ivory – Advanced Testing Methods

Advanced testing methods are those undertaken in specialized laboratories, when simple tests cannot give an accurate or detailed enough answer. Such tests are of value to officials in their attempt to combat the illegal ivory trade, and to museum curators, historians and archaeologists. However, there is no single laboratory test that will answer all questions; indeed many tests can only give indications of a single factor that may aid the identification of the material being tested.

In museums it can be helpful to know the age of an item, and whether it has been repaired. To archaeologists and historians it is useful to know whether very degraded items from excavations are of ivory or bone, which can appear very similar, as that tells a lot about how people lived in past times. Equally, it is interesting to know whether items are made of elephant ivory from Asia or from Africa, as this can help to unravel the complexities of the old trade routes.

Research is always ongoing, and new tests are also being developed, although the quick and easy test to determine on the spot whether ivory is illegal or not still proves elusive. One of the problems with ivory is that it is a mixture of mineral and organic material, and many tests are only suitable for one or the other.

Many laboratory tests are labelled as 'destructive'. This means that a sample of the material must be taken from the item to be tested. The sample is usually very small, in the region of 50–200 mg, but removing it inevitably leaves a mark. While it may be possible to find an area from which a sample can be removed from a large carving – for example, from the base – without altering the integrity of the piece, it is more problematical with small artefacts, inlay or mounted items. In the case of beads, samples can usually be taken from inside the drill hole.

For many of the tests it is necessary to have a database of results from previous tests of specimens of known provenance, with which the new results can be compared.

Many tests used in gemmological laboratories are of use in differentiating between ivory and various simulants or fakes, but not between the different ivories. Two of the more common tests are FTIR and Raman spectroscopies, which take readings based on how chemical elements react to various wavelengths in the electro-magnetic spectrum, and compare the results against known data.

There are numerous tests which concentrate in one way or another on the chemical composition of the material being tested in order to determine the type of ivory. As all dentines are basically made up of the same elements the tests look at the differences in the trace elements or the various isotopes present, using forms of Trace Element Analysis and Isotope Analysis. They work by analysing what an animal has absorbed from its food, as the old adage 'You are what you eat' is true.

Two more advanced but very useful tests are carbon 14 dating, and DNA testing. These can tell the age of the ivory and the species from which it came. A relatively new form of testing to differentiate between animal species analyses protein sequences, an example of which is ZooMS.

FTIR Spectroscopy

Fourier Transform Infra Red (FTIR) spectroscopy is an extremely useful laboratory test, which measures the reaction of a specimen to different frequencies in the infra red part of the electro-magnetic spectrum. The results are calibrated by a computer, which gives a reading of the specimen's absorption of the light at each wavelength. The reading is compared to the earlier test results from known samples in order to analyse it.

There are two methods used. Either a sample is taken from the specimen and pulverized to give a reading by transmitted light, or the reading can be taken straight off the specimen by reflected light. The first is by nature 'destructive', but the second is not always possible, and depends upon the size of the specimen.

FTIR is excellent for differentiating between ivories and their simulants, and also for identifying the type of simulant used, but differences among the various ivories can be so small that the results are not obvious and would need to be read by someone very experienced in the field.

FTIR can detect the water content of a sample, so theoretically it can differentiate between elephant and mammoth ivories, as the latter loses its water content and becomes more mineralized. However, the mineralization of mammoth ivory would depend somewhat upon the conditions in which a tusk has been preserved, and similarly an elephant tusk could be old and dry. A further problem is that the outside of a tusk is more mineralized than the inside. There could therefore be too many variables for this to be a totally reliable test.

Raman Spectroscopy

Raman spectroscopy is somewhat similar to FTIR, but uses monochromatic light from a laser beam. It measures the way particles in the beam (photons) are scattered and altered when hitting the specimen. As with FTIR, it is necessary to have readings from a known sample against which the results can be compared. The test is non-destructive.

Raman *can* differentiate between the various ivories but, because they are very similar, results tend to overlap. It is not therefore a totally reliable test for this purpose.

Trace Element Analysis

Trace element analysis measures the minute traces of elements that have been absorbed by an animal from its food or surroundings.

There are various tests, one of which is ICP-MS (Inductively Coupled Plasma Mass Spectrometry).

By analysing the trace elements in the chemical composition of a sample taken from a specimen and identifying what the animal has ingested and absorbed, it can be possible to determine where it lived.

The test is destructive as a small sample must be removed from the specimen.

The test is not suitable for tusks that have been buried (for example, mammoth tusks) as they pick up trace elements from the surrounding soil. However, it can be of use in determining from which area an animal came. A drawback is that animals move around, so a sample may not give a full picture as the trace elements will then vary from the outside to the inside of a tusk or tooth. A sample from a carving would also not give a complete picture.

A further drawback is that it is, of course, necessary to know what is present in any given area that will result in the animal absorbing the specific trace elements – for example, trees or grasses.

Another test is LIBS, or Laser-Induced Breakdown Spectroscopy. It measures the component parts of a sample and their concentrations. In ivory the main components are the same, but the trace elements that are detected can differentiate between, for example, elephant and mammoth ivories.

LIBS targets a minute area of a sample with a strong laser pulse and vaporizes it. A second laser pulse excites the vaporized particles and, as they cool down, they produce an electro-magnetic emission from which the readings are taken.

As with other tests there can be an overlap, and also much depends upon possible contamination from storage or burial for thousands of years, where, for example, mammoth tusks have been found, and which part of a tusk is being tested.

Isotope Analysis

Isotope analysis is also a way of examining the elements in an ivory sample, and from them deducing where the animal has been and what it has eaten, but instead of looking at the trace elements it looks at the isotopes.

Most of the readings are taken from the collagen, and special attention is given the nitrogen, carbon, hydrogen and oxygen isotopes. For example, the nitrogen isotopes give indications of soil types and the hydrogen and oxygen ones are linked to water. Other isotopes can give further indications of geographical provenance.

It is a 'destructive' test because a small sample must be taken. It would be an unnecessarily complicated test to use when determining whether an item is ivory or a simulant as there are simpler tests that are more efficient – although it can differentiate between specimens such as bone and ivory, which in archaeological specimens can be difficult.

Carbon 14 Testing

It is possible to determine the age of ivory by the use of carbon 14 analysis. This test is expensive, and can only be carried out in a few, specialized laboratories. Most importantly, though, it cannot tell when an ivory carving was made, but only when the ivory-bearing animal died.

Ivory has at times been stockpiled by places such as monasteries or cathedrals, as it was regarded as a prized asset on a par with gems, gold and other valuables. A clear advantage of stockpiling was that the material was available to be used in their workshops whenever required. Thus, the year when the animal died is not necessarily an accurate date for the carving. Also, there have been occasions when ivory was difficult to obtain and so old items were re-carved, sometimes hundreds of years later.

It is not surprising that experts cannot always agree on the age of a carving based on style and iconography alone. On such occasions, and where it is of importance to know the age, carbon 14 can be very useful as a guideline. In some instances the ivory itself proves to be younger than the carving was reckoned to be. In others a carving can look old, yet there is something 'not quite right' about it, and through testing by carbon 14 it may be discovered that the piece consists of ivory of two different ages. The most likely reason for this would be that it is an old carving, but with a later repair.

The results of the test can throw up some surprises, as was the case when the Cloisters Cross was tested. This mediaeval carving, thought to have been made in Bury St Edmunds in England, is now in the Cloisters Museum, part of the Metropolitan Museum in New York. It was believed to have been carved somewhere between around the years 1050 and 1200, but when the carbon 14 test was carried out the ivory proved to be up to five hundred years older than that, so had clearly come from a place where they had a stockpiled collection – as would most probably have been the case at the abbey in Bury St Edmunds.

Carbon 14 dating (often written as C14 or 14C), is a test whereby we can ascertain the approximate age of organic materials from a couple of years old to about 60,000 years old. Thus we can test ivory carvings, none of which have so far been dated to more than 40,000 years old. We can also get readings from mammoth or mastodon remains that are less than 60,000 years old as they have been buried in the permafrost and frozen rather than fossilized, so that they still retain some of the collagen content in their teeth, tusks and bones. Their C14 is in the organic collagen, not in the hydroxyapatite (*see* Section 2.1 What is Ivory?).

Living organisms contain carbon. Most of it is stable, but a minute percentage – the isotope called carbon 14 or 'radiocarbon' – is radioactive and unstable. Plants or animals absorb C14 throughout their lives by photosynthesis or through the food chain, but cease to do so when they die. The C14 isotope then starts to degrade.

Dating is based on the degradation of the C14 present in a sample. Knowing that it has a 'half-life' of 5,730 years, i.e. the length of time it takes for the carbon 14 content to halve (a process which keeps repeating so that it halves again in another 5,730 years, and so forth), it is possible to judge the age of the item.

The calculation of age according to the C14 readings is not totally straightforward, however, as various events over the years have influenced the carbon presence in the atmosphere, thus affecting the results. For example, in the 1950s when atomic bombs were being tested, it was realized that lots of C14 was created, with the consequence that natural levels were raised until the test ban treaties came into force. At its peak the amount doubled. Allowances are therefore made for that period and readings taken today are calibrated to allow for the distortion – indeed, the fact that there was a boost of carbon in the atmosphere at a given date has enabled the testing of more recent items to be more accurate.

It is important to know from which species the ivory being tested has come. For instance, walrus ivory from a thousand years ago will give a different reading to elephant ivory of the same age. Again, this must be calibrated as, if not taken into account, a walrus ivory item may be thought to be three hundred or so years older than it actually is.

In younger items C14 readings are comparatively accurate. Going back a thousand years they are less accurate, and the results are usually regarded as being correct to within plus or minus about seventy-five years.

C14 tests can also be easily contaminated. A sample must be taken from slightly inside an item as any oils, waxes, resins or colours on the surface may alter the test results. Depending on the preservation of the collagen, a sample of 200–500 mg of powdered ivory is needed. Although a relatively small amount, this means that the C14 test is a destructive one, and is therefore only used on small or rare objects when it is deemed absolutely necessary.

By proving that confiscated ivory is fresh and therefore comes from an animal that died very recently, C14 testing can be of help in the control of the illegal ivory trade. However, as every tusk in a haul would have to be tested to give an accurate representation it would be extremely expensive and time-consuming and is therefore seldom used.

DNA Testing

DNA testing will not tell us the age of an ivory carving, but it may help to determine its provenance, although only by telling us from which species the ivory was taken.

Other tests can tell the differences between different animals – for example, whale or walrus ivory, and indeed also between ivory and bone or a polymer – but so far only DNA can differentiate with complete accuracy between African and Asian elephant ivory. It has been described as being 'more finely tuned' than, for example, FTIR or Raman. However, problems can be encountered even with DNA.

DNA is extracted from cells, but there is no cellular structure in dentine. Although it is produced by cells called odontoblasts in the jaw of the elephant, only small remnants of these cells remain in the dentinal tubules which are present in ivory (*see* Section 2.1 What is Ivory?). Also, DNA degrades with time, so what can be obtained from ivory – especially old ivory – is very fragmented DNA. This can be difficult to compare with examples of DNA taken from living creatures.

Furthermore, when testing very old ivories there may be no extant examples with which to compare it. This is the case with certain Asian elephant sub-species which are now extinct. Therefore, results would be labelled as either 'African' or 'non-African'. Of course, any elephant that is non-African is Asian, but without a sample from the exact species it is not possible to give it an exact name.

DNA has the drawback that it is a slightly destructive test. It is a very sensitive test, and a small sample of material must be taken from inside the object to be tested. It is not possible to take a sample from the surface as that will be contaminated by its surroundings and give inaccurate readings. Although it may be possible, in a large carving, to find an unobtrusive area where a tiny hole can be drilled from which to take a sample, a small carving or a piece of jewellery poses a much bigger problem. For this reason DNA testing is often deemed unsuitable.

Other drawbacks with DNA testing are that it is expensive, it cannot be done *in situ* but must be done in a totally clean and sterile environment in a specialized laboratory, and results are not immediate but can take some days.

All this notwithstanding, DNA is invaluable in the determination of whether new, confiscated ivory is Asian or African. In such samples the DNA is far less fragmented and it is much less problematical to get good results. Laboratories have been developed in various places such as Bangkok, Thailand, where a small, government-run laboratory has been set up specifically for this purpose. There, confiscated raw tusks are tested to discover their provenance. The laboratory has been established because Thailand is a known hub on the illegal ivory trade routes.

DNA testing is also immensely valuable in cases of ivories being sold in shops as legal 'mammoth ivory' when they are, in fact, illegal elephant ivory carvings. However, there is a reluctance to do a lot of testing because of the cost, so efforts concentrate mostly on illegal imports of raw ivory.

ZooMS

Zooarchaeology by Mass Spectrometry (ZooMS) is an analysis of animal protein sequences. It identifies molecular 'bar codes' or 'fingerprints' – that is, the pattern unique to each genus. It is especially useful in identifying archaeological bone specimens – hence the name.

Collagen is extracted from a specimen, cut by an enzyme (usually trypsin) into peptide fragments, and the mass of each fragment is measured using time-of-flight mass spectrometry. (Mass spectrometry works by ionizing the peptides to generate charged molecules and measuring their time-of-flight to estimate their 'mass'.) Test results are compared against a database of masses from previously analysed samples.

Mapping protein sequences is more costly than DNA analysis, but once sequences are known, measuring the peptide mass fingerprint is cheap and has other certain advantages. The test is non-destructive as the sample needed is so microscopic that it does not mark the specimen, and it can be taken *in situ* and from an artefact of any size, large or small. A small area of the artefact to be tested is rubbed with a soft eraser, and surface molecules removed electrostatically onto crumbs of eraser. From these they can be extracted and tested later in a laboratory. Thus it would be possible to test the inlay in a piece of furniture to discover whether it is ivory or bone, or to test a masterpiece in a museum, without removing it from the venue.

Another big advantage is that protein survives approximately ten times longer than DNA, so an item can be successfully tested for protein sequences after its DNA has decayed and fragmented and is no longer viable. This is especially useful for degraded archaeological specimens. The protein sequences can also survive various forms of processing, so, for example, they can identify leathers after they have been tanned and the DNA destroyed. This is of use in the policing of the illegal trade in wildlife parts.

ZooMS can distinguish between elephant ivory and most other ivories, also between elephant and mastodon, but not between elephant and mammoth ivories as, to be successfully tested, the species must have an evolutionary divergence of at least five million years. The mastodon line diverged at about that time, but the mammoth line diverged a little later. ZooMS is also unable to differentiate between Asian and African elephant ivories.

2.5

Determining the Age and Origins of Ivory Carvings

Ivory loses its lustre and becomes harder as it dries out. In archaeological finds, ivory items may be so desiccated that it is not possible to tell whether the material is indeed ivory, or whether it is bone or antler. In such cases laboratory tests are required.

In ivories that are several hundreds of years old rather than thousands, and which can be seen to be ivory, it can be very difficult to determine the age of a carving, or its origin. As explained in Section 2.4, scientific methods of testing can cause damage to an item – albeit minimal damage – so they are always avoided, if possible. Instead, reliance is put on visual evaluation, and judgement based on experience.

Provenance can get more and more vague with time. Items change hands, diptychs are broken up, tabernacles are taken apart. Later restorations may have been badly executed, and incorrect parts of items reassembled or replaced. Polychromy is an unreliable guide as there is very little original colour on old ivories today.

To estimate the age of a carving, the style, workmanship and iconography of the piece are studied and, where possible, compared with items of known provenance and age (for example, other ivories, or sculptures in stone or wood). Illustrated manuscripts and paintings may also act as guidelines, as inconsistency in style for the item's presumed age – such as dress or hairdo – may be a sign that it is not authentic. In past times it was customary to depict people wearing the latest fashions. A piece will also be examined for signs of artificial ageing such as surface colouration, improbable cracking, wear and repair, and the general condition of the ivory material.

In later years workshops may have kept detailed records of what they produced, but in past centuries this was seldom the case. In the instances where they were kept, the records are today of great value and help when judging other carvings of unknown provenance.

For hundreds of years the master craftsmen have travelled. For example, a young man in mediaeval times might train in Paris – which was possibly the most famous of carving centres, and with a style admired and copied all across Europe. Having proved himself an accomplished craftsman, he might then find work in Germany, before moving on to Denmark to continue his craft there. The carvers went wherever there was work, and a master carver could also be requested by a foreign royal court, or an abbey, or even a nobleman, to come and work for them. There are styles that are typical of some schools of work, but the influences amongst the countries were strong and the distinctions become blurred. It is for this reason that many items in museums have labels such as 'Probably made in Italy' rather than giving Italy as the definite location.

Good examples of the problems faced were those encountered when attempting to date two of the three large, double-sided, mediaeval crucifixes that have survived to this day. The Ferdinand and Sancha Cross, named after the King and Queen who commissioned it for the basilica of San Isidoro de León in the mid-eleventh century, is made of elephant ivory and its provenance is known. Today it can be seen in the National Archaeological Museum in Madrid, Spain.

The Gunhild Cross is in the National Museum in Copenhagen, Denmark. It is made of walrus ivory and is about 30 x 20 cm in size, and the Christ figure is missing (see Fig. 3.12). The cross is inscribed as having been made for 'the daughter of King Svend'. For some time this was presumed to be Svend Estridsøn, who died in 1076, on which basis the cross should have dated from around that time. Further research showed that the style of the Latin script was later, so the 'Svend' referred to is most probably Svend Grathe, who reigned from 1146–57, which would mean that the cross is younger than previously thought. The style of the cross has also puzzled researchers, with some saying that it is English in style while others say German – although it was actually carved in Denmark.

The third mediaeval cross can today be seen at the Cloisters Museum (part of the Metropolitan Museum) in New York. It is also carved in walrus ivory, and is also missing the Christ figure. It measures 58 cm x 36 cm, and is made up of five parts which lock together. There are many theories about the provenance of the Cloisters Cross, one of the most convincing being that it was made for the Abbey at Bury St Edmunds, England, in the twelfth century. It then disappeared for eight hundred years. The cross appeared again in the middle of the twentieth century when it was offered for sale by a dealer of slightly dubious background, who refused to give any details of its recent history. Several museums were interested, and it was finally bought by the Metropolitan Museum in New York in 1963. There is no doubt that it is mediaeval, nor that it was carved in England, but more than that cannot be said with certainty. It remains a fascinating historical art mystery.

A different problem arises when trying to determine the age of tribal carvings, as, for example, the ivories from the former Belgian Congo. Such items are often labelled by museums as 'Acquired' at a certain date, rather than when they were made. The reason is that the carvings were classical, standard shapes or figures, made in exactly the same way and style for centuries, often in connection with special ceremonies. They all look alike and their age cannot be judged by sight, but would need to be tested in a laboratory. Even so, the result may not be exact.

In the East the problem of dating carvings can be exacerbated by the fact that it has not been considered 'forgery' to copy a master craftsman's work; indeed, this was the way in which apprentices learnt their craft. Signatures could also run in families and be passed down the generations. Furthermore, the scenes and styles used in the Far East are often very classical, with little change in subject matter or how it was interpreted for hundreds of years. Therefore, unless a piece is actually dated, it may be virtually impossible to tell from its style when it was made.

In Japan, carved ivory items were usually signed, but interpreting the signature and putting a date on the carving is still difficult as, not only did successive generations use the same or similar names, but apprentices could also sign a piece incorporating the name in their own. To make matters even more complicated, a carver could use either his own name, or his name as an artist, which might be completely different, and names could be written in different styles of character as well.

Some museums have kept strict inventories of all the items that have come into their possession. These records – especially the inventories of some older museums – are of immense value in the dating of unknown ivories as it is possible to compare a piece of unknown origin and age with one that is on the museum's list.

When it is imperative to try to discover the age and provenance of a piece, and the material from which it is made, testing can be undertaken in a specialized laboratory (*see* Section 2.4). However, testing is expensive, and many museums do not have the spare funds to carry out such investigations. Furthermore, many of the tests are 'destructive'. That is not to say that an artefact will be destroyed, but for some tests it is necessary to take a small sample of material from the item in order to carry out the test. It is usually possible to find an area where a small scraping or tiny hole would not be noticeable, but it is not something undertaken lightly on a very old and valuable masterpiece. Furthermore, while it may be all right to take a small sample from a large carving, it is far more difficult to do so from something small. One more drawback is that the tests can give misleading results.

In addition to determining the age of an artefact, it can also be useful to see if there have been repairs made to it. A very old carving may have had a missing piece replaced, and this could have been done so long ago that it is not immediately apparent. A simple test like viewing the object under ultraviolet light may show up repairs, and further testing would then be indicated.

Ivory carvings from many places originally had a little colour added to enhance the carving. This is true, for example, of many ivories from the Gothic period. Today, they still appear to be coloured but the pigment is very seldom the original, and is most likely to have been replaced in the nineteenth century. Again, laboratory testing is necessary to analyse the pigments as the old ones used are very different from those of the nineteenth century.

Elephant tusks have been traded worldwide for many centuries (*see* Section 3.1 Raw Ivory Trade Routes). It is usually sufficient to know that an item is made from elephant ivory, and seldom necessary to know whether it is Asian or African, but on occasion this knowledge can help to give an indication of the object's provenance. While not completely foolproof, a simple guideline states that if a piece measures more than 11 cm across at its widest part, it must be made from African elephant ivory, as the Asian tusks were too small to accommodate these dimensions. This rule is, of course, of no use with smaller items.

Today there is an area where it can be very necessary to judge the age and provenance of an item: the illegal trade in ivory. Most commonly it is whole elephant tusks that have been seized following a raid, but individual carvings can also be in question. Ivory can be artificially 'aged' to give it the appearance of being old. Methods used include heating it to cause it to crack, or staining it. The best form of identification of such 'fakes' is by careful examination (*see* Section 2.2 Ivory Substitutes, Fakes and Simulants *and* Section 2.7 Working Ivory).

There are various tests that can be undertaken in a specialized laboratory. For example, there is carbon 14 dating, DNA, and others (*see* Section 2.4 Testing Ivory – Advanced Testing Methods). However, so far there is no quick, easy and inexpensive method suitable for use by the authorities at places like ports of entrance or in shops. The search for a good system continues.

2.6

Care and Damage

Ivory is a delicate material. Although it can survive for thousands of years given the right conditions, in the wrong ones it can deteriorate quite quickly. Care should therefore be taken both in use and in storage.

Humidity and Heat

Ivory is a hygroscopic material: that is to say, ivory can absorb moisture and release it again. It naturally contains a small amount of moisture, and this should not be allowed to dry out as it will cause cracking. Storing ivory in low humidity (below 35 per cent) may also cause it to dry out and crack. Absorbing too much moisture, and then drying out – especially if it dries out too fast – will cause it to crack or warp. The ideal storage conditions are in an environment of stable relative humidity and temperature, at between 40–55 per cent relative humidity, and between 16°C and 20°C.

Even if it is dirty, ivory should never be washed as this will make it too wet. If it has been soaked it is very difficult to dry successfully.

Ivory objects should never be displayed in places likely to get too hot or too dry, which can cause cracking and warping. Examples are on a mantelpiece over a fire (where an object can also become discoloured by smoke and lose its high polish), on a sunny windowsill, or in a glass case lit by spotlights. Ivory should not be kept in the dark, but too much UV light can also cause damage, especially to old artefacts which are a little more vulnerable.

Some ivories are more prone to cracking than others. Elephant ivory is especially susceptible, and it tends to crack in two directions: in concentric rings following the growth structure of the tusk, and along the tusk in longitudinal direction, as can be seen in many mediaeval carvings. Walrus ivory can crack along the length of the tusk and also between the primary and secondary dentine.

When once ivory warps, or cracks have appeared, there is very little that can be done to mend it.

Breakage

Although a whole tusk is a solid and tough material, carved ivory is far less so. The finer the carving, the more brittle and vulnerable the material becomes. If dropped, a carving is likely to both break and crack.

▲ **Fig. 2.28** Warped base of an ivory box, which cannot be reattached.

◄ **Fig. 2.29** Broken *okimono*. As it was comprised of several pieces, it has also fallen apart at the joins.

Surface Damage

Ivory is a relatively soft material, and can be scratched if it comes into contact with abrasive surfaces. Ivory jewellery, such as beads and bangles, is especially susceptible as it can knock or brush against hard surfaces during wear.

Extra care should therefore be taken in any attempt to clean an ivory artefact, since even apparently soft materials may be slightly abrasive to ivory.

► **Fig. 2.33** Hairbrush back damaged by chemicals. It has also become too dry and cracked.

Ivory can also be attacked by chemicals on the surface, which not only dry it out but can leave clear marks. Harsh chemicals can mark it very quickly, but leaving drops of perfume on the surface will also slowly damage it.

Cracks

Ivory that has cracked cannot be put back together again. Gluing ivory in one place will cause tension elsewhere and another crack will form. If a piece is noticeably cracked, the gap can be filled in with epoxy resin.

▲ **Fig. 2.34** Sperm whale ivory carving of a seal which has been badly damaged by drying out, causing excessive cracking.

An alternative is to use a decorative and contrasting filler to make a feature of the crack. While this would clearly not be an option for an old and valuable ivory artefact, it has been used to effect in, for example, mammoth ivory jewellery, where otherwise poor quality pieces of material can be utilized (*see* Fig. 1.9).

Surface Colour and Patina

Ivories vary in colour, and also in how their colours alter with age. Very old ivories may become much darker with time and appear dry and somewhat woody.

When kept in the dark, ivory usually darkens, and can go almost a yellow colour. This can often be seen on the base of an ivory carving, where the rest of the item is still pale, but the base – which is seldom subjected to light –

▼ ► **Figs 2.32** & **2.33** Ivory *netsuke*. Note that the base (*left*), which never gets any light, has darkened.

darkens. The lid of a piano with ivory keys should not be kept closed, as the keys will turn yellow.

Ivory can be lightened again by bleaching, but this is a drastic measure and not advised. Any chemicals used are highly likely to damage the ivory by removing natural moisture and drying it out. Bleaching an item in sunlight will also dry it and cause it to crack. It is better to leave an object that has darkened and regard the colour as patina. One hundred years or so ago, fashion dictated white ivory. Subsequent enthusiastic scrubbing with soap and water has done much damage to old ivories.

Fig. 2.34 Ivory box used for cosmetics. The staining is more apparent in the area of the ivory taken from near the pulp cavity of the tusk.

Ivory can look dry, with a loss of surface lustre. It can even feel dry to the touch and lacking in its usual, slightly resinous feel. This does not necessarily happen only when an item is very old, but can also happen to relatively young pieces, and there can be various reasons for the condition. For example, the ivory may not have had a very high polish from the start, as may be seen in some African carvings. In the case of display items the surface may have been incorrectly or over-zealously cleaned at some stage. Washing a piece of ivory may reduce its surface lustre. Even constant dusting with a cloth that is not very soft can gradually abrade the surface and reduce its lustre.

As already stated, chemicals can damage ivory, marking the surface and ruining the lustre.

Staining

If ivory comes into constant contact with oils, tobacco, tea, make-up, and similar substances, it can stain. Thus, for example, boxes used for the storage of face powder may become pink inside.

For many years a pair of very large elephant tusks stood by the doors of the Natural History Museum in London, where passers-by could not resist stroking the beautiful, soft surfaces. After some time it was realized that the tusks were turning dark brown as a result of the constant stroking, which transferred minute amounts of natural oils from the hands of the people, every time the tusks were touched. Sadly, there was no alternative but to remove the tusks in order to preserve them.

Staining usually penetrates the ivory, so a simple re-polish of the surface will not necessarily remove the colour. However, if the surface is painted with a water-based pigment the colour can be removed, although it may leave a trace. Extra care must therefore be taken with items such as painted miniatures.

Cleaning

Any cleaning is best undertaken by professionals, as the incorrect use of chemicals, or even water, can damage ivory beyond repair.

There is an old idea that ivory can be washed in milk. This presumably derives from the assumption that the milk will clean it while, at the same time, replacing some of the 'oil' that may have been washed out. This is not, strictly speaking, true, as any added oil can attract dirt and so exacerbate the problem. Also, ivory should only be slightly dampened on the surface, and should never get very wet. There are some enzymes which are used to clean ivory without damaging it, two of which are to be found in cow's milk. It is therefore difficult to say categorically that milk should never be used. Nonetheless, amateur use of any liquid, including milk, cannot be condoned, especially as any damage caused may be irreparable.

Among the professionals, such as museum conservators, there are varying methods used for cleaning ivory (*see below*).

Ivory items should not be indiscriminately dusted, as incorrect handling can so easily cause damage and breakages. If an item is dusty, extremely gentle brushing with a very soft, dry artist's paintbrush should suffice.

Conservation and Repair of Ivories

Museums have ivory conservators, whose job it is to repair items and attempt to restore them to their former glory. They use the knowledge that has been gleaned over decades of work, in order to get the best result. In some instances this is a very big task. Following the First Gulf War many of the Nimrud ivories disappeared from the Baghdad Museum, believed to have been looted (*see* Section 3.5 Egypt and the Middle East). However, it turned out that they had been removed to safety and were brought to light some years after the war had ended. Unfortunately, when they were unpacked it was discovered that the place in which they had been stored had flooded in the interim, and most of the ivories were very badly water-damaged. The job of drying out 3,000-year-old carvings, neutralizing anything that could damage them further, and then cleaning them was beyond the knowledge available locally, so expert help was brought in from abroad and the local conservators were taught the specialized art. It took many years of work to restore the ivories. Even so, many will never be the same again.

While it is possible to clean ivories, remove dirt, and neutralize any damaging environment, attempts are seldom made to bring old, discoloured ivories back to their original creamy colour, and they are left with the patina that they have acquired. Few, if any, of the Nimrud ivories – especially those mentioned above – are now of a creamy colour.

Another problem was encountered with the Bagram ivories, discovered during excavations near Kabul in Afghanistan. They were found early in the twentieth century, when knowledge of how to treat them was limited. The ivories were found to be in a very friable state and, not knowing how to lift them from the ground without breaking them, archaeologists came up with the idea of pouring warm, liquid gelatine over them to act as a stabilizer. Unfortunately, the gelatine hardened and trapped dirt around the objects, and also attracted more dirt. It also shrank, causing the ivories to flake. In further efforts to protect them, the ivories were given more coats of gelatine, were mended with successive applications of glue, and stuck to paper. When some of the samples were finally correctly restored in the early twenty-first century, all the incorrect substances had to be removed and were replaced with reversible resins that did not damage the ivory.

When regular cleaning of ivories is undertaken to remove everyday accumulated dust and grime, conservators use specialized liquids in minute amounts. Some museums use artificial forms of various enzymes, including salivary amylase, which in its natural form is present in human saliva (*see also* Cleaning, *above*). Only the smallest amount of the liquid is used on a very soft piece of cotton, wrapped around a small stick which has been whittled to a point. It is an exceedingly painstaking procedure and, because of the delicacy of such work on carved items of ivory, where any intricate carving can easily be snapped off, it is strongly advised to let only an expert undertake the task.

Some museum conservators prefer to use a mixture of ethanol and water rather than use enzymes, but again the utmost caution is advised, and such work should only be undertaken by experts.

Cracks cannot usually be repaired as the material has warped somewhat, thereby causing the crack. Forcing a crack to close would simply cause the ivory to crack elsewhere. Furthermore, cracks usually contain some dust, which is difficult to remove, so a dark line would still be visible even if the crack were completely closed.

Ivory that has been broken *can* be repaired, though not invisibly. As very specialized glues are used and immense care has to be taken because of ivory's hygroscopic nature, such repairs should also only be carried out by experts.

Many old carvings are missing parts. For example, a carving of a Madonna may be missing an arm or hand because it was carved from a separate piece of ivory and that has become detached and lost. If a new piece is added, the restoration may involve giving the mended area a light coating of pigment in order to match it to the surrounding ivory and hide the repair. Usually the new area of ivory needs to be darkened, but on occasion white is added to the surface to lighten it. All repairs to museum pieces must be reversible, therefore water-based pigment is used as it can be removed again.

Another common problem occurs in old ivory boxes or book covers, which have metal hinges or studs. Those made of brass leach the copper content into the surrounding ivory, turning it green. Silver can stain the ivory grey to black, and iron turns it brown. Conservators draw the stains out using a variety of poultices, but it is seldom possible to remove the stain completely.

Another puzzle is why ivories develop different patinas as they age. Some darken; others do not. It is known that most ivory darkens when kept in the dark, as can be seen on the base of figures that have stood for a long time (*see* Fig 2.32), or piano keys where the lid of the piano has always been kept closed.

As the ivory ages, various other factors clearly affect the colour, but research is still needed to fully understand why. Some centuries-old figure carvings may have darker areas on the clothes, possibly left by the residue of paint, or of glue used to attach gold leaf. Other carvings that have been handled a lot are affected by oils from skin, darkening the surface. Some very old carvings turn almost brown, or even grey, yet others do not. This is a consequence – at least in part – of the conditions in which the ivories have been kept, but as yet we do not know why certain conditions affect only some ivories and not others.

Conservation is a relatively new discipline, and methods are still being developed or perfected today. In times past it was largely a matter of trial and error and there have been several occasions where, in trying to conserve the ivory, it was damaged further. Fortunately, today, conservation is better controlled and knowledge and understanding of what is involved is much greater.

Note: The author and publishers accept no responsibility for ivories damaged by attempting any of the treatments mentioned above, and would always advise seeking expert help for repairs and similar.

Fig. 2.35 Box with a metal hinge which has stained the ivory. The value of this box is not great enough to merit trying to remove the stain.

2.7

Working Ivory

Cutting the Tusk or Tooth

The first step in working ivory has always been to cut the tusk or tooth into the required shape and size. Great care was taken with this part of the procedure, as ivory was expensive and waste had to be kept to a minimum.

The way that a tusk or tooth was cut depended largely on what sort of finished item was intended. Some tusks were limiting by their shape (for example, boars' tusks) and others by the fact that they were hollow (for example, narwhal). Walrus tusk limited the craftsman because of the two different types of dentine present.

Elephant ivory was the most versatile, being the largest, with more solid and homogenous content than the other types. Even so, an elephant tusk had to be sectioned according to the end product.

A large carving in the round (for example, a Madonna and Child) viewed from the base will be seen to be hollow for part of its length, and the sway of the figures follows the curve of the tusk, thus maximizing the use of the tusk and avoiding too many joins. Any part that protruded from the figure, such as an outstretched arm, was likely to have been made separately and attached. Maximizing the use of the ivory available dictated the typical shape of these carvings and the stance of the Madonna's figure.

Sliced, cut panels and plaques, which were to be carved in relief and used on furniture or for attaching to wooden boxes, were usually about 1 cm thick and cut along the tusk rather than across it.

The largest slice of ivory could be obtained by cutting the tusk at a slight angle, just above the pulp cavity. Depending on the angle, it gave an oblong

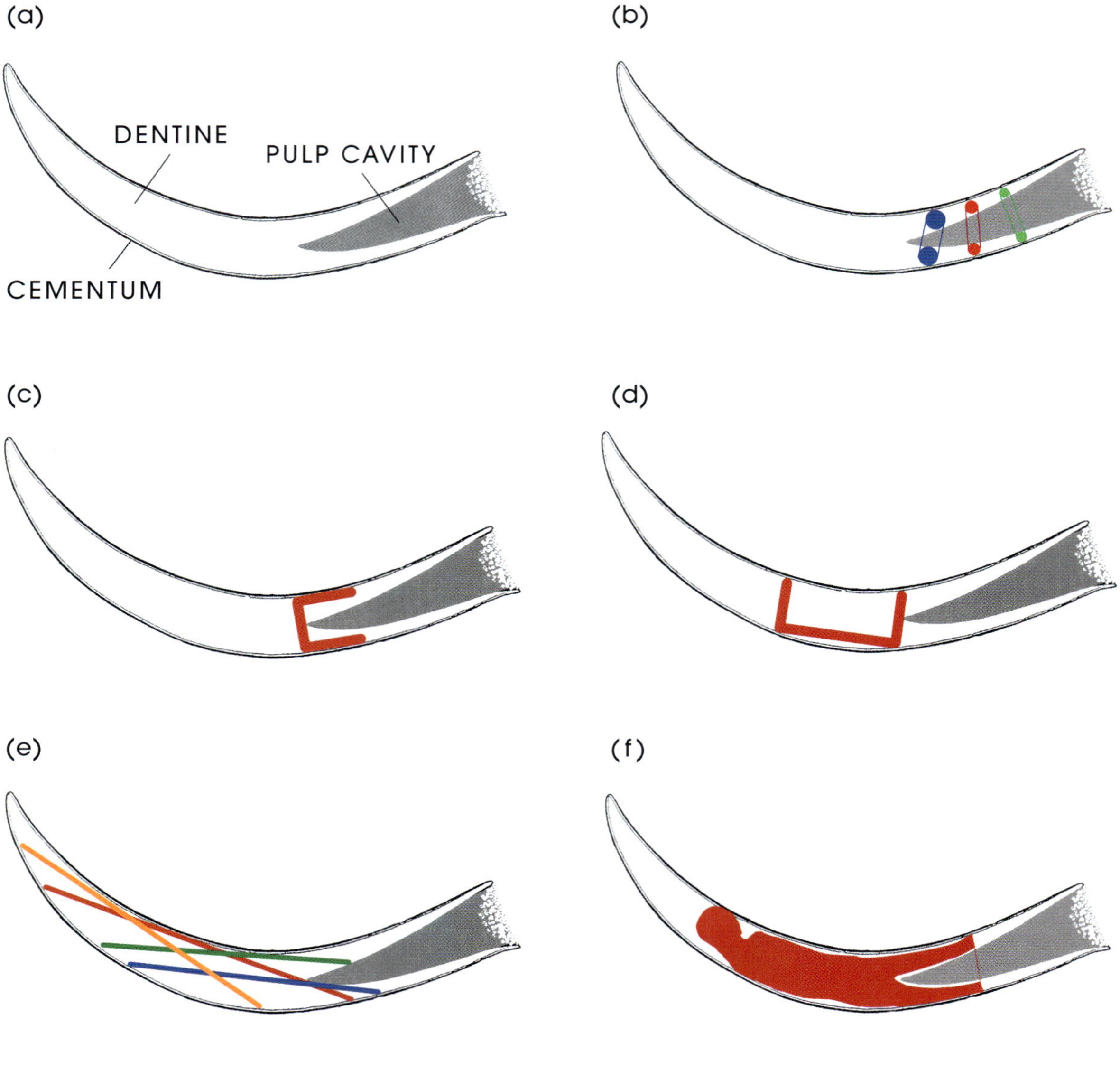

Fig. 2.36 Cutting an elephant tusk: (a) basic diagram of tusk cut in half lengthways; (b) bangles cut around the pulp cavity; (c) box cut in one piece, around and beyond the pulp cavity; (d) box cut in one piece from beyond the pulp cavity; (e) the areas from which the longest possible slices can be cut (note that the intersecting arc pattern would not appear in cross-section but would instead appear 'stretched'. *See also* Fig. 1.24.); (f) the position of a figure carved in one piece (note the 'sway' of the body and the base cut around the pulp cavity).

or an oval shape and, when looking for the intersecting arc pattern in the ivory, it is neither on the flat surfaces nor at right angles to them (as would usually be the case with items such as boxes), but at the edge of some of the surfaces. Quite often it is possible to see the cone-in-cone pattern of the ivory in a very large slice. Items such as hand mirrors, which needed extra length for a handle, were always cut at an angle.

In some cases the ivory carver attempted to avoid cutting near the pulp cavity because, although that could give the largest slice from any given tusk, the dentine near to the pulp cavity tended to be softer, more fragile, and would possibly even have a slightly different colour compared to the rest of the tusk.

The sides of round boxes were cut from the section of tusk containing the pulp cavity, and a base was attached afterwards. The base was usually a straight slice of ivory taken in longitudinal section of the tusk, not in cross-section, which again maximized use of the material. In mediaeval times the bases of some pyxides (small, round, lidded boxes) were cut from one solid piece of tusk without joins, which was an extraordinarily extravagant use of ivory (*see* Fig. 3.8, the Cordoban pyxis). The lids were similarly cut from a single piece.

Hardness of Ivory

The hardness of gem materials is measured on Mohs' scale, which is a scale of 1–10 based on the ability of harder materials to scratch the surface of softer ones. Specific minerals are used as standards. Thus talc is taken as the standard soft material at number 1, as it can be scratched by gypsum at number 2. Diamond is the hardest at number 10.

Although often quoted in connection with ivory, the actual numbers have little relevance except to demonstrate that ivory is a relatively soft material, roughly on a par with amber, jet and pure gold.

Enamel would be about 6 on Mohs' scale, but when working ivory any remaining enamel is always removed, so its hardness is only of consequence because it can be difficult to remove.

The hardness of the cementum and dentine varies, although cementum contains more collagen and is, in theory, the softer material. However, the hardness of dentine decreases the nearer to the pulp cavity it is measured, and cementum gets harder as it ages, so the cementum on the outside of a tusk may well be harder than the dentine underneath. On Mohs' scale both dentine and cementum are quoted as being between 2¼ and 3. In practice that is of little consequence as, for aesthetic purposes, the cementum is also usually removed before working the ivory. Exceptions are sperm whale and orca ivory, which have very thick cementum layers that are usually an integral part of the finished item.

To the ivory worker, the comparative hardness of different ivories and their constituent parts is more important than any numbers. Together with size and shape, these factors determine for which purpose the different ivories are most suitable, and how they will be worked.

From a carver's perspective, the hardness and suitability for carving are as follows:

- African elephant ivory may be a little harder than Asian elephant ivory, but it is preferred because the tusks are much larger.
- Walrus ivory primary dentine is like elephant ivory.
- Walrus ivory secondary dentine is best avoided as it can crack or splinter.
- Hippo ivory is the hardest of all ivories.
- Narwhal ivory is more brittle than other ivories.
- Mammoth ivory is harder than elephant ivory.
- Fossilized walrus ivory is harder than mammoth ivory, but it is more stable and less likely to crack.
- The cementum layer of sperm whale ivory is much harder to carve than the dentine.

Carving

It is believed that the oldest ivory carvings found were made about 40,000 years ago. Clearly the original methods used were extremely primitive, and would have consisted of somehow cutting a small piece of ivory from a large tusk, before shaping it with bits of sharp stone (*see* Section 3.2 Prehistoric Times).

With the advent of metal tools, a process developed that has not changed very much over the millennia:

- Before any work was done, a sketch of the finished item was made. In some workshops it was also customary to make a model in clay or wax.
- Having chosen a seasoned tooth or tusk, any enamel remaining was usually removed by abrasion. Sometimes it was removed by the application of acid, or by heating it and flaking it off with a chisel.
- The appropriate size for the design was calculated, and cut from the tusk using a saw. It was important to work out exactly how to cut the ivory so as to make the best use of the precious material and to waste as little as possible.
- In cases where the ivory had not been ordered by the carver, but simply supplied to him by a benefactor or a dealer (e.g. in India), the

carver chose his design by the size of the piece of ivory he had been given.

- The design was drawn on the surface of the piece, which was then roughly shaped using saws and chisels.
- If a whole tusk was used (as, for example, in oliphants), the sections for the various patterns could be determined by tying a string tightly around the tusk. This gave a line at right angles to the axis of the tusk. A string wound around the tusk was also used to determine the position of the spiral pattern in some African carved tusks.
- The use of various chisels, gouges, files and burins enabled more detailed work to be done. A piercing saw might also be used for carving right through the material for intricate and fine detail.
- Finer tools were used for fine work, eventually down to a sharp needle if necessary.
- In some instances and in some places, it was customary to moisten the surface of the ivory before executing very fine work in order to make it a little softer and less brittle.
- The piece was polished with a variety of abrasive materials.

The processes were speeded up when once electricity was used to power some of the tools. Today, in places where ivory is still worked, there are carvers who still only employ hand-tools, with the possible exception of sawing the ivory into the desired size and rough shape which can now be done on electric band saws and rotary saws. Any small pieces that are cut off are kept to be used in the making of small items such as jewellery. The craftsmen using hand-tools tend to be those working individually, while the electric tools are used in the factories in the Far East for mass-produced items.

On tusks and teeth where there was only a small amount of enamel remaining, it was usually abraded manually or ground away mechanically. The enamel which covered hippo teeth was also usually removed by abrasion, although it might be removed by the application of acid. This was a tricky process and risked damaging the dentine underneath. Another method that has been used was heating and fast cooling, which enabled the carver to scale off the enamel, though again there was a risk of damaging the dentine and causing it to crack.

The basic cutting and fashioning methods have varied little from country to country, but the materials used for polishing depended on the country in which the ivory was carved. For example, in Europe they used abrasive powders followed finally by leather, while in the Arctic seal skins, in the Far East shark skin, and in China wet bamboo leaves might be used instead. In India the carvers dampened the surface of the finished item before polishing. They then used rough fish scales as an abrasive, and finished with common chalk to give the final shine.

Nowadays drawings and models are still made, and a rough outline of the figure is cut. Hand carving is done using a variety of chisels, burins and files. The tools used for the fine work have varying shapes and functions – some have flat tips, others pointed, and some have a sharpened curved surface, all depending on what is to be achieved. The surface is smoothed by scraping away the rough areas, and polishing with successively finer grades of abrasive papers, before a final shine is given with a solution containing wax. This is not to coat the ivory, but to bring out its natural lustre.

When electric tools are used, rotary files in various sizes are employed, the smallest of which are akin the dentists' tools. The danger with using electric tools is their speed. A craftsman has more control when using the old-fashioned, slow methods, and a mistake made when carving ivory cannot be undone, but only disguised.

FORMS OF CARVING

There are many styles of ivory carving. The following is some of the terminology used.

- **In-the-round** refers to completely three-dimensional items (for example, carved statuettes) where a piece of ivory is carved on all sides.
- **Relief** carving consists of three main types. When an item is carved in relief the back of the piece is flat and the design or picture is on one side. It is known as 'bas-relief' or 'low relief', when it is a shallow form where the ivory is not cut very deep and, although three-dimensional, the carving has a rather flat appearance. Cameos were carved in low relief. 'Haute relief' or 'high relief', has much more depth to the carving and includes some undercutting, making parts of the design truly three-dimensional. The third form is 'pierced relief' or 'openwork', where holes in the ivory plaque are cut through with a saw and finished with finer tools, possibly to give a lattice-effect background. This was also used in small carved picture 'silhouettes'. A special name was given to the lattice-style, finely pierced relief that was much used by the carvers in Dieppe: *mosaïque*. This is also sometimes termed *ajouré*, as it is reminiscent of the style of lace known as *ajour*.
- **Undercutting** occurs where the carver has cut all the way around or behind a section of what would otherwise be a relief carving. For example, an animal may be depicted on a plaque with the legs nearer

the back attached to the background, but the legs further forward carved all the way round. This can also be included in 'high relief' carving.

- *Bachiru* is a Japanese name for carving into a piece of dyed ivory to reveal the natural, pale colour beneath. The incised areas were sometimes dyed contrasting colours. It was used in the eighth century in Japan and was also popular in China during the Tang Dynasty.

Fig. 2.37 Elephant ivory carving in 'low relief'. Diameter 7 cm.

Ivory is very durable and strong, and can be carved into amazingly fine strips without breaking. An example is the rigging on models of sailing ships. It is most easily carved when it is 'green' (that is, very fresh) as ivory dries slightly when stored. Ships' rigging, the very fine lattice-work in some carvings from China, or the lace-like *mosaïque* work from Dieppe were not possible to execute on dry ivory as it would have become a little brittle. The carvers had to dampen the ivory before attempting such work, unless it was extremely fresh. The very fresh ivory was said to have a consistency similar to very hard cheese to cut, but to be much tougher. Although intricate carving is carried out in mammoth ivory today, it is not as delicate as that done in times past in elephant ivory, indicating that the ancient ivory is simply too hard and dry.

When using older ivory, or when repairing an old piece, some ivory carvers first dampen the surface of ivory with pure water, as this softens it slightly. Similarly, thin slivers of ivory edging or inlay need to be moistened a little or they will snap.

Turning

Lathes have existed for many centuries; indeed there are indications that they were already being used in Egypt in

Fig. 2.38 Ivory card case cover, photographed on a black background which can be discerned in the thinly carved areas which are translucent. Height 7.75 cm.

the Middle Kingdom (2055–1650 BC), but fashioning ivory specifically by turning it on a lathe became more popular in the sixteenth century. The ivory was mounted on a wooden bench, and turned by a foot pedal. A series of cogs and levers determined how a fixed blade cut into the ivory. The blades that were used varied in shape, depending on the profile to be cut.

The simplest turned forms were small pots or boxes, but large, ultra-elaborate celebration lidded cups or goblets measuring 60 cm tall could also be produced on a lathe. The designs did not have to be completely cylindrical – they could have wavy edges, scalloped profiles, open corkscrew spiral stems, and all manner of complicated patterns, determined by the settings on the lathe and, of course, the skill of the craftsman. The results always had rounded edges rather than sharp, cornered ones, but they were not necessarily round in profile.

Centres for the craft were mostly in Germany, and included Nuremberg and Dresden, although fine work was also executed elsewhere. The ivory used was almost exclusively elephant tusk.

In the sixteenth and seventeenth centuries it was a hobby in many royal courts to dabble in producing things on a lathe, and it became a pastime of both kings and queens. Among the turning enthusiasts were Count Franz of Erbach-Erbach, Peter the Great of Russia, and Queen Sophie Magdalena of Denmark, all of whom were quite accomplished in the art.

Several hundreds of years earlier, in Al Andalus, blocks of elephant ivory tusks were shaped on a lathe, preparatory to carving them. It was the custom in about AD 900 to carve pyxides from a solid piece of tusk, rather than to join pieces together (*see* Cutting the Tooth or Tusk, *above*, and Fig. 2.36c). A lathe was used to give the base of the pyxis a perfectly round shape, so that the lid could fit exactly. The insides were also hollowed out on a lathe.

◄ **Fig. 2.39** Lidded cup of turned elephant ivory, made for display only. Note the various profiles used. Nuremberg, early seventeenth century. Height approx. 50 cm.

Slicing in Thin Sheets

'Peeling' was the word usually used to describe the cutting of very thin sheets of ivory, probably because the action was reminiscent of peeling an apple. It is a misnomer, however, because ivory does not peel – it is sawn.

Jean-Henri Pape was a successful piano-maker living in Paris in the early nineteenth century. He was mechanically minded, and soon modified and changed some of the accepted methods of piano construction. He invented a machine similar to a lathe but with one long blade, that could cut wooden logs or elephant tusks in one continuous piece so that it unrolled, resulting in extremely thin and large sheets of material. This enabled him to veneer his pianos in ivory. It is said that he could make a sheet almost 3 m long from one tusk. Thin sheets of ivory could be used as veneer on other furniture, but only if the wood was very pale and the glue used was colourless, because the ivory was so thin – supposedly less than 2 mm – that it was almost transparent. Thicker slices would not unroll flat, but would crack. The machine was sometimes used for making the large sheets of ivory used as a 'canvas' for paintings.

The more common method of slicing ivory was to use a special veneer saw and cut the material from large tusks at the point and angle where the largest slices could be obtained. Using this method slices of about 3 mm thickness were obtained. At that thickness the ivory was less transparent.

Ivory Combinations

Ivory complements other materials, not least other organics (those of plant or animal origin). In the sixteenth and seventeenth centuries it was often combined with amber in the making of shrines or caskets, the edges being strengthened with ivory, and it was added as finials or plaques for decoration.

Ivory was also often combined with tortoiseshell to add strength to the edges and corners or the keyholes on boxes or tea caddies, where the tortoiseshell alone would have been prone to chipping.

As marquetry on furniture, ivory was combined with materials such as tortoiseshell (*see* Fig. 3.16, cabinet with ivory, dyed bone and tortoiseshell marquetry by Pierre Gole), and by the early twentieth century it became fashionable to combine it with shagreen in boxes, furniture, games pieces and so forth (*see* Fig. 3.20, backgammon pieces).

There were a few famous ivory master carvers who liked to combine it with other materials. One was Simon Troger (1683–1768). He was born in Germany, but studied in Italy before going to live and work in Austria, and finally returning to Germany. He was known for his figures combining ivory with dark wood, where the ivory depicted skin, while the wood depicted clothes, horses, or whatever else was in the scene. His figures always had glass eyes.

In Art Deco times it was not unusual to carve statues combining ivory and bronze, again the ivory depicted the skin while the bronze depicted the clothes. Such figures were usually mounted on marble plinths. One of the famous craftsmen of the era was Demetre Chiparus, a Romanian who studied and lived in Paris. Favourite subjects were dancers and, following the discovery of Tutankhamen's tomb, Egyptian themes.

Undoubtedly the most spectacular use of ivory in combination with another material was the larger than life-size chryselephantine statues, made for the temples in Ancient Greece. Although ivory and other metals have, in later times, also been referred to as chryselephantine, the true combination is ivory and gold. As in the other examples mentioned, ivory depicted the skin, and gold depicted the draperies, hair, and everything else. The statues were made by laying sheets of carved ivory over a wooden base, and thin, beaten gold over the rest of the statue. Some statues had semi-precious stones or glass added for details such as the eyes.

The most famous ancient chryselephantine statues were those of the goddess Athena in the Parthenon in Athens, which reputedly stood thirteen metres tall, and that of Zeus in the temple at Olympia which was slightly shorter at twelve metres tall – but he was seated. The statue of Zeus was considered one of the Seven Wonders of the Ancient World.

There were several chryselephantine statues made at the time, all of them showing off the wealth of Ancient Greece, although periodically the gold was stripped off and melted down to be used as currency. None of the ancient statues remain today.

Incising and Etching

Over the years many and varied items have been incised and etched, preparatory to having pigment rubbed into the lines. Japanese *netsuke* and *okimono* often had the pattern enhanced by incised lines, emphasizing, for example, hair or fur. Furniture was also subjected to this treatment, where small, inlaid pieces of ivory were made to look more intricate with some added lines, and simple designs or lines might also be added to any ivory object as part of the design. For purely practical purposes sundials depended upon the lines drawn and etched as they enabled the user to tell the time.

Anything to be incised had to be well polished, as a high polish withstood the pigment better and so gave much clearer lines in the incised areas. A rough surface absorbed the colours and gave a messy result.

The most usual method of incising letters, numbers or patterns into ivory was to scratch the surface with a sharp point and then rub lampblack (soot), ivory black or ink into the incised area, before wiping away the surplus. This was one of the methods used for making scrimshaw (*see also* Section 3.11 The Americas and the Caribbean).

A more complicated method was to cover the item in a wax solution: the pattern or design was then etched into the ivory through the wax, exposing the incised ivory beneath. Concentrated sulphuric acid was poured onto the surface and the exposed parts turned black. After rinsing thoroughly in water, the wax was dissolved away with turpentine.

Instead of using sulphuric acid, silver nitrate could be coated onto the surface in the dark, and exposed to the light. Being photosensitive, the silver nitrate turned black. The wax was then removed in the same way. Using gold chloride in place of silver nitrate gave a purplish-red colour (*see also* Colouring *below*).

Bleaching

Today we would seldom try to bleach a piece of ivory, preferring to leave it with its patina of slight yellow colour, and knowing that bleaching can cause a lot of damage. But in times past it was quite commonplace to bleach ivory if it became a little discoloured.

One method used was to steep the ivory object in turpentine for several hours, boil it in soapy water for three hours, and finally cool it in cold water before drying it away from direct sunlight. Other methods used a variety of different solutions, including hydrogen peroxide, and a mixture of urine and lime.

Colouring

On occasion the natural, creamy colour of ivory was altered. This could be for a special effect, usually to make it look older than it actually was, which was common when attempting to fake an old item. The colour can be altered by various processes, such as heating, smoking, or staining with liquids to give a darker colour.

From ancient times ivory has been coloured according to the fashion of the time, or the culture of the country, or for more practical purposes such as to tell games pieces apart, as in chess sets, draughts, or billiard balls (*see* Fig. 2.45). Ivory was sometimes dyed green when used in marquetry, to contrast with white ivory and brown woods or other materials.

In mediaeval times it was popular to enhance a carving with a little pigment. For example, a carving of the Madonna and Child might have a blue border added to her cloak to emphasize the line, or the background in a carved relief might be coloured in to give contrast (*see* Fig. 3.14 The Virgin and Child from St Chapelle).

Generally, however, colour was used sparingly on ivories, and in later centuries it was sometimes considered incorrect to have any colour at all, so it was removed from many of the existing pieces. In the nineteenth century many ivories were re-coloured, often incorrectly. There was no sense of careful restoration of an old artefact to return it to its original state, but instead the work was considered to be an improvement on the original, frequently carried out by dealers to get a better price for the object. Today it is not uncommon for conservators to find more than one layer of pigment on an item.

Before the invention of aniline dyes in the middle of the nineteenth century, various pigments were used for colouring. In mediaeval times colours were limited and most of them were toxic, so working with them was not without danger. For instance, a yellow pigment usually contained arsenic, and reds contained mercury. Most of the pigments were also extremely expensive. The commonest colours used were blue, green, dark brown and various reds. Some common constituents of these colours were:

- **Blue:** lapis lazuli, or sometimes azurite, both of which are minerals.
- **Green:** verdigris, obtained by exposing copper to acidic fumes.
- **Red:** cinnabar (mercuric sulphide). Also called 'vermillion'.
- **Brown:** a concoction obtained by boiling certain woods.

Finally, some less harmful, organic, dyes were discovered. For instance, turmeric (a spice) was a good yellow dye and replaced the one made partly from arsenic, while cochineal – which was made from insects – gave an excellent red. Indeed, by the time the first billiard balls were being made, these were probably the dyes being used. Green colour was obtained by treating the ivory twice, first turning it yellow using turmeric, followed by treating it with blue indigo, made from the plant *Indigofera tinctoria*. This combination gave a green colour.

Before any colour could be applied to the surface, the ivory had to be mordanted, that is steeped in a solution which would enable the subsequent dye to fix. Mordants varied according to the dye chosen, and included tin-salt solutions, hydrochloric acid, or aqua fortis (nitric acid).

Entire surfaces could also be coloured by immersing the ivory, in the dark, in silver nitrate or gold chloride. These gave a light-sensitive coating which turned black or purplish-red upon exposure to light (*see also* Incising and Etching, *above*).

Any area which was not to be coloured by one of these methods was coated in wax prior to treatment, and often a thin wax coating was also given after the dyeing process to prevent the colour fading in sunlight.

Different processes were used for miniature paintings on ivory (*see* Section 2.8 A Few Worked Items).

When a pattern was incised into the surface of ivory and the picture or design was coloured to bring it out, various pigments were used.

- Lampblack, or Indian ink were commonly used by scrimshanders.
- In India *lac*, from the lac insect, was dyed (usually black) and rubbed into the patterns incised on furniture.
- In Japan each studio had its recipe for colouring its *netsuke* or *okimono*, and these were closely guarded secrets.
- In ancient times some ivories were deeply incised – almost carved intaglio – and the cavities filled with coloured wax to give contrast.

Gold Leaf

At various times over the millennia it has been the habit to add gold leaf to ivory carvings. Seldom was the whole carving covered, but areas would be highlighted, often in conjunction with paint.

The purer the gold, the softer it is, and the thinner the sheets that can be made from it. The material used in ancient times was considerably thicker than gold leaf as we know it today, which is so thin as to be almost transparent. By the Middle Ages it was being made in thinner sheets, but they were still about twice the thickness of the modern material.

The gilding process involved painting the surfaces which were to be gilded with a fish glue, and then carefully laying the thin leaves of gold onto the surface and gently pressing it into place, usually with a soft brush. Surplus gold could be simply brushed away.

Few old carvings still display their original gilding, but slight staining from the glue used, and minute traces of the gold itself, may remain as an indication. (*See* Fig. 3.10 A Sicilian casket with traces of gilding and pigments; *also* Fig. 3.14 The Virgin and Child from St Chapelle.)

Some carvings had polychrome and gold added in the nineteenth century. It is usually possible to test any pigments to see if they are old or not, but it can be more difficult to determine the age of the gilding.

▲ **Fig. 2.40** Elephant ivory carving of the frigate, *Norwegian Lion*, by Jacob Jensen Nordmand, Denmark, 1654. The sails have been partially bent into shape by softening the ivory.

Softening

Over the millennia, various methods for softening ivory have been suggested. It was said that the Ancient Egyptians were skilled in the work, and the Greek historian, Plutarch, wrote in about AD 100 that ivory could be softened in beer.

By 1837, when the fifth edition of *The Cabinet Maker's Guide* was published, the method advocated was to soak ivory in 'best vinegar' to which chopped mandrake (a plant belonging to the nightshade family) had been added.

Some old writings maintain that ivory could be softened to the extent that it could be pressed into metal moulds, and re-hardened by immersing it in common salt. If this was true, the method has now been lost and is a mystery. However, it would seem rather fanciful, and possibly best regarded with the same distrust as Plutarch's statement that ivory could be heated and moulded in the same way as cow horn. (Horn is a keratinous material which can be moulded with heat and pressure: ivory is not.)

Nonetheless, it is acknowledged that thin sheets of ivory can be softened and bent – for example, to shape the billowing sails on carved ships, where simply carving their shape in ivory would be a very extravagant use of the material. Further, ships' rigging was also carved by first softening the ivory very slightly to make it less brittle.

While it *is* possible to soften and bend ivory using water alone, it is thought that a slightly acidic solution was used to aid the process. Aqua fortis (nitric acid) in a dilute solution seems to have been the favourite liquid in which to immerse the ivory sheets for a few days, after which they were pressed over wooden moulds until they dried and hardened. Both the bending and the drying processes had to be undertaken with great care as the ivory was susceptible to cracking. The use of dilute phosphoric acid was another old suggestion, but it apparently removed some of the ivory's opacity.

It has been suggested that the ivory coverings on the chryselephantine statues of Ancient Greece also consisted of softened ivory sheets over carved wooden carcasses. Although it is not possible to prove, it seems likely that the ivory on the arms and legs was softened and moulded, but that on the faces was carved in order to obtain the necessary detail.

Inlay and Veneer

'Inlay' is, in fact, a fairly loose term referring to attaching pieces of one material to a carcass of another material. As with other forms of ivory work, it was often carried out in family-run workshops, and the craft passed down from one generation to the next.

- **Intarsia** is a more precise term for one form of inlay. This involves cutting a pattern out of the base or carcass, and filling it with a contrasting material cut to exactly the same shape and size. The most common base is wood, and the most common 'inlays' – which are glued in place – are contrasting woods, ivory and mother-of-pearl. This is the method used often in the Indian furniture made for the European market from the seventeenth century and later.
- **Marquetry** is a more sophisticated method that followed intarsia. This also uses contrasting materials. However, in marquetry the whole surface is made up of small pieces, following a template, which are cut to fit together like a jigsaw puzzle. They are usually assembled before being mounted on the carcass which, again, is usually wood. They may be glued first onto a piece of paper and the whole thing is glued on to the carcass, or they may be slid into place without a paper – all depending on the skill of the worker.
- ***Khatamkari*** is a form of marquetry which is known to have been used in Iran about seven hundred years ago, although some records suggest that it was developed there much earlier. The finished result appears to be an extremely fine mosaic with repeating geometric patterns. The method has been used throughout the Middle East, and also in India where it was called *sadeli* work. It involves shaping very slender, straight rods of different woods, metals, and bone or ivory, giving them a variety of cornered profiles. The most intricate is a star shape, often used in Iran, while diamond shapes were more common in India. The rods are taken one by one and bound and glued together in such a way as to form a geometric pattern when the whole block is viewed in cross-section. The block is cut into slices about 1.5 mm thick, and the slices are glued onto a carcass (usually wood), in a geometric pattern. By gently heating the slices it is possible to bend them slightly and apply them to curved surfaces. *Khatamkari* work is still made in parts of Iran, and *sadeli* is made in India. Today, the white areas are usually made of plastic.

▲ **Fig. 2.41** Glove box with *sadeli* work marquetry, made in India, 1855.

- **Veneer** is a thin sheet that covers the entire surface of the carcass and is held in place with glue.
- **Banding** is a form of marquetry or intarsia. It may be a simple strip of a different material attached to the edge of a piece of wooden furniture or box, or it may be a more complicated strip of geometric design, incorporating different materials. Ivory banding was often used in the form of a single, narrow strip of ivory edging a tortoiseshell-veneered box (for example, a tea caddy); this narrow piece of ivory would then protect the tortoiseshell from flaking at the edges.
- ***Shibayama*** work refers to the method of inlaying materials with other materials, but not presenting a flat finished surface. When ivory is used as the core material, the work is very delicate and intricate. Indentations in the ivory are filled with small pieces of mother-of-pearl, tortoiseshell, coral, and so forth, cut to the exact size, and arranged to make a picture. The inlay is glued into place with fish glue, and the whole surface polished but allowing the inlays to stand a little proud of the ivory background, adding to the pictorial effect (*see* Section 3.9 Japan).

▼ **Fig. 2.42** *Shibayama*-style studs. Elephant ivory inlaid with mother-of-pearl and other materials.

All items that are inlaid using any of the above methods are polished after assembly.

Traditionally, the glue used for these various methods was a plant resin, or a type of fish glue called 'white-fish glue' or 'isinglass'. Isinglass was originally made in Russia using the swim bladder of sturgeon, but it was discovered in England that it could be made using cod instead. The fish material was dissolved in diluted alcohol –

preferably gin – and then bottled and subsequently boiled. The resulting opaque white jelly could be diluted with more alcohol if necessary. It was melted for use by placing the bottle into hot water. The alcohol content prevented the glue from decomposing. White-fish glue was colourless so it did not discolour the ivory, and could glue ivory to wood, or ivory to ivory.

In India, craftsmen sometimes used dark-coloured plant resin as glue when making intarsia. On ebony it covered any small discrepancies in size of the piece of inlaid ivory, while on paler woods it gave a black outline to the ivory, adding to the effect.

Larger pieces of ivory (for example, plaques set into pieces of furniture) were held in place with ivory pegs and possibly secured with glue as well. This was also the method used for ivory-covered boxes. If an item was made entirely of plaques of ivory with no wooden carcass underneath – as, for example, the mediaeval Sicilian boxes – the plaques were usually held together with metal wire.

Shavings and Chips: Paint, Polish and Medicine

Ivory has always been very expensive, so none of it was wasted. Off-cuts, chips and shavings were collected and bagged, and were sold. Larger pieces might be used to make inlay, but even shavings and 'dust' were put to use.

Although it is considered to be the most pure and beautiful creamy colour, ivory could be made into the blackest of black pigments: ivory black. This was done by taking ivory chippings or scrapings and incinerating them in a closed vessel over a slow-burning fire. The same method can be used on bone, but the resulting colour is slightly less dense.

There were many uses for ivory black. Apart from artists' paint, it was also used in making jewellery in the sixteenth and seventeenth centuries, when it became fashionable to mount diamonds in an enclosed setting and back them with black

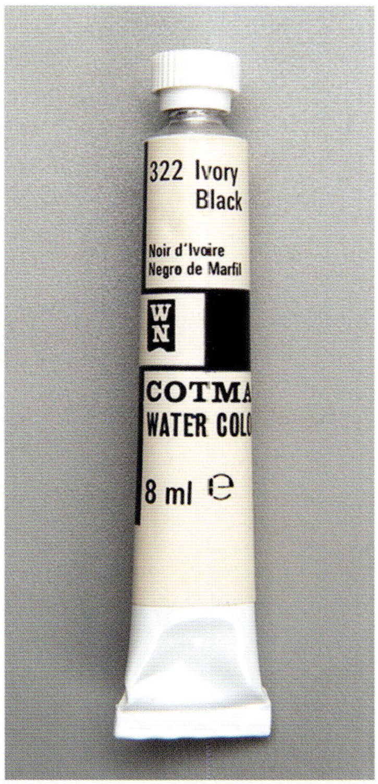

Fig. 2.43 Tube of ivory black watercolour paint.

foil or colour of some sort. One of the colours used for this purpose was ivory black, mixed with resin.

Ivory black was once one of the constituents of Indian ink, although lamp-black or soot were also used. Today ivory black may no longer be made using ivory, and charred bone or synthetic dyes have largely taken over.

Mixed with an appropriate resin, ivory dust has also been used as a soft polish.

There were also medicinal uses for ivory powder and shavings. One recipe using ivory shavings was for 'ivory jelly'. Recommended by Mrs Beeton in her famous book, *Household Management*, ivory dust could be boiled together with herbs and spices, and sieved. The resulting liquid was mixed with sherry and sugar, and left to set. It turned to jelly because of the collagen content of the ivory. It was noted that the ivory dust 'should be obtained from first class grocers', and that the resulting jelly was very nourishing.

Ingesting ivory for health reasons was not a new idea. In AD 77 Pliny the Elder wrote in his *Natural History* that powdered ivory could cure whitlows, and for a long time powdered ivory mixed with honey was a treatment for acne. Powdered ivory mixed with goat's blood was said to be a cure for gall-stones.

Ivory burned to a white ash and ingested was said to prevent miscarriages, cure constipation, resist malignant fevers and cure vapours and fits, and also to kill worms in children.

In the Far East ivory has been used for medicinal purposes for many centuries. It is said that its use continued in some parts until the mid-twentieth century, while in parts of Africa and India ivory dust was given to cattle to increase their milk yield.

Note: The author and the publishers accept no responsibility for damage caused to ivory objects if attempting any of the above treatments, and would always advise seeking the help of an expert.

2.8

A Few Worked Items

Piano Keys

At the end of the nineteenth century and beginning of the twentieth, the production of piano keys in New England in the United States (*see* 3.11 The Americas and the Caribbean), and also in England, UK, reached its peak. Piano keys were made of wood, and veneered with slivers of ivory. In one method of production the tusks were cut into a series of blocks, some short for covering the front of the keys, and others 10 cm long. The long ones were then divided into shorter blocks for the 'heads' – that is, the part that would be touched by the pianist – and some were reduced in width to make the 'tails' that fitted between the black keys. The blocks were then sliced into slivers about 1–1.5 mm thick, washed,

Fig. 2.44 Piano keyboard with ivory-covered keys.

dried, and bleached in weak hydrogen peroxide for a couple of days because traditionally the keys had to be very white. (Those for harmoniums and organs were not bleached.) The slivers were mounted on a large wooden block according to what was required, and polished. Finally, the block was cut into appropriately sized pieces to make a keyboard. It is estimated that in around 1900 America was producing over 300,000 pianos a year, all with ivory-covered keys.

Later it was discovered that the new material, 'plastic', could replace ivory. Pianists generally acknowledged, however, that because of its hygroscopic nature, ivory gave the best feel to a piano, and that plastics were a little too slippery. In spite of advances in the production of polymers, some pianists still hold this opinion.

Billiard Balls

As with piano keys, the main manufacturers of billiard balls were the United States of America and England. Billiard balls were also made elsewhere – for example, in Russia, where mammoth ivory was used for the production. However, most ivory billiard balls were made from elephant ivory.

Not all tusks were suitable for making into billiard balls. Asian elephant ivory was popular, and the best tusks for the purpose were fairly straight and not too large, as it was important to carve the balls from as near to the centre of the tusk as possible, so that the ivory dried evenly. Carving the balls from the centre of a very large African tusk would have been wasteful. It was normal to obtain only three or four balls from one tusk.

The tusk was cut into suitable chunks, which were then cut roughly spherical. On a lathe, one side of the sphere was shaped into a perfect hemisphere,

Fig. 2.45 Ivory billiard balls.

before the piece was turned around, the smooth-cut side inserted into a wooden chuck, and the other half cut into a hemisphere, resulting in a complete sphere. Further turning on the lathe smoothed out any imperfections.

The ivory balls were left to season for five years. During this time they dried slightly, and any unevenness became apparent. They were turned again in order to ensure that they were perfectly round and would roll straight before being polished. Some were also stained, the colours depending on the game for which they were intended.

In spite of seasoning the ivory for five years, a change in atmosphere could cause a reaction in the ivory, and the balls often had to be replaced because they ceased to run smoothly, or because they cracked. It was very expensive to constantly purchase new ivory billiard balls; furthermore, the number of elephants killed for the purpose finally began to be realized. As a result, with the advent of the new synthetic plastics towards the end of the nineteenth century, the billiard manufacturer Phelan and Collender in America announced a competition to find an artificial substitute for ivory billiard balls. The prize was $10,000.

Although substitutes were eventually found it was a long time before they were widely used, as it was difficult to produce a substance that rolled with the same momentum as ivory. Also, when the balls hit each other, ivory was deemed to make a more satisfying sound.

Scrimshaw

Scrimshaw is the name given to the artistic handiwork carried out by whaling boat crew members on any part of the whales, orcas, seals or walruses that they caught or traded with the locals where they were hunting. It covers anything from carved whalebone objects such as set squares – which were used on board ship – to magnificent engraved walrus tusks. Typical items produced included ships' equipment such as tackle blocks, cleats, awl or chisel handles and so forth. These were usually carved in bone, leaving the ivory for utilitarian items that might also be decorative, such as carved swifts (wool winders), jagging wheels, sewing boxes, ditty boxes, bodkins, crochet hooks, candlesticks, board games and picture frames. Today the best-known of all the scrimshaw items are the purely decorative ones: the engraved whale teeth, and walrus tusks.

Scrimshaw was often made as a present for a wife or a sweetheart back home, who had to be left for months – and sometimes years – while the ships were at sea catching whales to produce oil. Not all the whale teeth reached home though, as many of them were traded for tobacco during the voyage.

The art of scrimshaw was universal. Wherever there were whaling boats, there were scrimshanders, whiling away the time on board when they were not working, and often singing sea shanties and playing music at the same time as scrimming.

It can be argued that the art of scrimshaw goes back to the Eskimos of the Dorset Culture who scratched pictures on bits of bone, but it is generally understood that, although the methods and the materials used may be similar, the term 'scrimshaw' denotes the whalers' art.

The production of scrimshaw was at its peak in the nineteenth century, when whaling ships were numerous. Many ships carried more crew than required to sail the ship, as the extra hands would be needed when a whale was caught and was subsequently rendered and reduced to oil. At one time it is said that there were more than seven hundred whaling ships ploughing the seas, the majority from the coast of New England in the United States.

The commercial whaling ban of 1986 virtually put an end to the art of scrimshaw. The trade in the ivory teeth of sperm whales and orcas, as well as the tusks of walruses, was already illegal thanks to the Marine Mammal Protection Act passed by Congress in the USA in 1972. This stated that it was illegal to trade in the teeth if they were harvested after 1972 – though they could still be given as presents. In fact, scrimshaw was already a dying art owing to the earlier discovery that mineral oil could be extracted from the ground, which meant that oil did not have to come from whales. The advent of mechanized vessels, with more efficient methods of catching and killing the animals had also given the sailors fewer hours to fill in between catches, so there was little time to practice the art of scrimshaw.

Not every sailor had an artistic streak or was very dextrous and the work produced by such people was quite crude, but some of the sailors were clearly talented and produced beautiful work. Even the very functional items were often quite intricately and imaginatively worked, especially when carved in ivory. Engraved baleen often depicted ladies in fashionable clothes – a subject suitable for decorating a busk. Some scrimshanders were able to produce designs freehand, but others traced theirs onto the baleen by putting an illustration from a magazine over it and pin-pricking the outline through the paper.

The same method could also be used on ivory, although ivory is a harder material. Fashionable ladies were also sometimes depicted on ivory, but more popular were pictures of sailing ships, especially the one on which the artist was living and working at the time. In America especially, patriotic images depicting flags, ships and historical scenes were also popular. A particular favourite was the Anglo-American War, which lasted from 1812 to 1815 and which was commonly known as 'The War of 1812'. Portraits of sailors, soldiers, wives or children were also produced, and quotations or dedications might be engraved on the back. Although a lot of engraved scrimshaw was dated, few objects were signed.

There are various methods of making ivory scrimshaw. Whichever way it is done, the first step is to polish the surface of the material and remove any imperfections. This cannot be done later as it would damage the image.

The traditional method of working is to polish the ivory to a good shine, first with shark skin and then with soft leather. Using a sharp point – classically, a sail needle – a pattern or picture is incised on the ivory either by scratching lines on the surface, or by pricking it. Only one tool is used for this. The depth of the line is not dependent upon the size of the tool used, but upon the pressure used. Light and shade areas are determined by the proximity of the lines. Each tooth or tusk must be tried for softness first to find out how hard to press the engraving tool, as no two teeth are alike. When the picture is finished, lamp-black (i.e. soot) mixed with whale blood, or black ink is rubbed over the surface with a cloth. Some sinks into the incised lines, and the rest is rubbed away. More can be added to the picture, and the inking process repeated.

Another method is to cover the whole surface with ink and let it dry before 'scratching away' the picture, so that it appears in negative. The surface is then re-inked, the fresh ink dissolving the dried ink and, as it is wiped over, the ink settles in the incised parts producing the picture in positive form – as it is intended to be seen. A final wipe with a damp cloth removes any excess ink. The advantage of this method is that it is easier to see and control the image as it is formed, because it is white on black rather than white on white.

Colours can be added to pictures. The incising methods are the same, and there is commonly a black outline made with the usual lampblack or ink. Colours are added afterwards, using a variety of dyes. The favourite of the whalers was water-colour paint, which was taken on board for that purpose. These paints did not cover the ivory in the same way as heavier pigments, and allowed the lustre of the ivory to show through.

Fig. 2.46
Unfinished scrimshaw on mammoth ivory, made by the negative to positive method.

When bone is used for engraved scrimshaw it has the disadvantage that it contains tiny channels (the Haversian system) into which the ink can run and give an unwanted effect. Ink and colours can also run on ivory, but this is less common, and is usually caused by the ivory not having been polished enough at the start.

Scrimshaw is still being produced today, although not, of course, on whaling ships. Today's scrimshanders use mammoth ivory, bone, antler, vegetable ivory, imitation ivory, and very occasionally a piece of old ivory such as a billiard ball, although in most countries reworked ivory cannot be sold for commercial gain. The designs used today are often pictures in roughly the same style as the old, traditional ones – however, the work produced is of a high standard as it is made by artists rather than sailors whiling away long hours on a ship.

Mammoth ivory is, today, the most common substitute for sperm whale or walrus ivory and, as it is basically the same material, it gives the best results. The outer surface, however, can be extremely hard as the material has lain buried for thousands of years. On the other hand the outer surface has been used successfully to add to the decoration (*see* Fig. 3.42).

Many modern scrimshanders use a single, pointed, metal burin, like the scrimshanders of old, though the modern burins may have a carbide tip. The old, laborious, polishing methods have given way to more modern methods, which use emery papers in increasingly fine grades, plus polishing compounds. Lampblack is no longer used as it is very messy and does not give the best results. Furthermore, some modern scrimshanders have abandoned the simple burin method and instead use electric tools such as a dentist might use. Although this may be faster and require less effort, it is tricky as it is extremely easy to make the incisions too deep when using electric tools.

Imitation ivories are not as satisfying to engrave and work in general, and they lack the feel and lustre of ivory. The ones most commonly used are polymer resins. Another ivory substitute is vegetable ivory, also called tagua nut or corozo nut. It is a natural substance but is, as the name suggests, a vegetable product and nothing like the traditional materials used of old (*see* Section 2.2 Ivory Substitutes, Fakes and Simulants).

Originally a gift from a sailor to his sweetheart back home, scrimshaw has now become a collector's item. Today original pieces – old and new – are sought after and fetch high prices. Inevitably, therefore, scrimshaw is also faked. So much imitation scrimshaw has been made that in America it has been given a name: 'fakeshaw'.

There are varying qualities of fake scrimshaw. It is usually mass-produced in imitation ivory, with the pictures drawn by machine. This sort of fake is made into small brooches and similar items.

Another method is to take a cast from an original tooth or tusk and make a mould. Soft resin moulds can be filled with a suitable, creamy coloured polymer resin and peeled off when the resin has set. They can take very fine detail and the moulds do not have to be made in two halves. The resulting 'tooth' has the picture ready incised, and just needs it to be blackened. By adding a stain to the outside of the 'tooth' it becomes even more convincing, as a real tooth would have become slightly discoloured through use during the animal's life. When left unmounted it is possible to see the inside of the 'tooth', and to see that it is not made of dentine. If the model is mounted and it is not possible to see the inside, it can be hard to tell whether it is real or fake, and detailed examination must be undertaken. An incorrect weight may give an indication of a plastic imitation. Plastics are lighter than ivory, which is a giveaway, so the resin used to imitate it has a filler added to give extra weight. Quite often such fakes end up feeling *too* heavy.

There is a thriving production of fake scrimshaw in China, where they favour two methods. The first is bone, carved in two halves to represent the front and back of the 'tooth', and glued together to look like a hollow tooth. If the join is decorated it can be almost unnoticeable. A second method is to use hippo ivory, either incisor teeth or the larger canines. Carved into the shape of a sperm whale tooth it has the correct weight and feel of ivory. The inside may be partially hollowed out to imitate a whale tooth's pulp cavity, leaving tell-tale cutting marks. The natural pulp cavity of a hippo tooth is different in shape from that of a whale tooth. As hippo ivory is harder than whale ivory the etching is unlikely to have smooth lines, but will be made using small, scratched lines. Finally, the iconography may be incorrect. For example, a picture of a Western person sometimes has slightly Oriental features.

Miniatures

Miniatures were painted on vellum, which is made from treated animal skin. In about 1705 a Venetian lady by the name of Rosalba Giovanna Carriera – who until then had decorated ivory snuff boxes – decided to use ivory instead of vellum for miniatures. The idea was a great success, not least because ivory did not have to be given a preparatory layer of paint, but, contrary to vellum, was itself considered the perfect skin colour. The rest of the picture could be built up on this.

The first ivory miniatures were painted on quite substantial slices of ivory, about 4 mm thick. By 1760 a method had been developed to cut the ivory wafer-thin – about 1 mm only – in fact, so thin that the slices were translucent

Fig. 2.47 Miniature painted on ivory. Probably starting with a photograph, it has been very poorly finished by hand-painting. Height 6 cm.

and could even be backed with a coloured paper or foil to give an extra glow to the finished painting.

The surface of the ivory tended to be a little too smooth to absorb the paint, so it was first prepared by roughening it a little with sandpaper or powdered pumice, soaked in vinegar to remove some of the natural oily feel, and occasionally left in the sun to bleach – although this last technique risked making it crack as it dried out. The pigments used to make the watercolour paint had to be very finely ground, and when used were mixed with gum arabic to help them to adhere to the ivory, sometimes with the addition of ox-gall or sugar. Oils were very seldom used as they defeated the purpose of the ivory by covering the material's 'perfect skin tone'.

Fig. 2.48 Reverse of ivory miniature. The ivory is so thin that the painting can be seen through it. Note also the visible structure in the ivory.

To give a slight shimmer to parts of the picture such as the sitter's hair, or possibly the leaves of a tree in the background, finely chopped hair could be glued to the surface. Finally, because the painted surface is very delicate and can easily be damaged, the miniatures were always covered with a convex glass and framed to be worn as jewellery, sometimes as a locket with a lock of hair set in the back as a *momento mori*, or to stand as a picture.

Fig. 2.49 The miniature in its ivory frame.

By the end of the eighteenth century the production of ivories had reached India, where English miniaturists set up business with the permission of the East India Company and taught the craft to local people.

Miniatures did not have to be very small. They could be as large as a tusk dictated. A piece of ivory, sliced at an angle across the tusk just above the pulp cavity, attains the maximum length available, and could produce an oval about 20 cm long. Further, in the 1830s, a Frenchman by the name of Pape discovered a method of 'peeling' an ivory tusk, which involved turning it on a lathe in ever-decreasing circles, cutting off a fine layer of the material along the way. The resulting thin sheet of ivory could easily be a metre long. It was then glued to a wooden base to stabilize it. However, the ivory tended to crack, as can be seen in surviving examples today.

▲ **Fig. 2.50** Puzzle ball in elephant ivory, consisting of five concentric spheres, carved from one piece of tusk.

The art of miniatures virtually died when photographs became a viable alternative, although there was one last twist. Fake miniatures were produced by painting a light-sensitive medium onto the ivory and making a contact print from a negative. This was then finished by hand with colours. In this way pictures of famous people or paintings could be mass-produced as miniatures at a far lower cost. Although most were fairly crudely executed, this was not true of all of them. Some were done so carefully that it can be extremely difficult to tell whether the image is a beautifully coloured photograph or a very fine painting. In these cases they have to be examined under high magnification, when it may be possible to see the silver nitrate crystals in the photographic emulsion.

As always, ivory miniatures have also been simulated in other materials, notably early plastics such as celluloid and casein. These materials were far less expensive than ivory, but the surface was not as good for painting as was ivory. Nonetheless they were painted, and also used for the more mass-produced, hand-coloured photographic type of work.

Puzzle Balls

A puzzle ball is a series of balls within balls. It consists of a large ball, containing a series of concentric balls of gradually diminishing size (anything from five to fifteen in number), which can all be independently turned within each other. In the finest carvings the outer ball is carved in high relief and all the balls inside are also covered in fine relief carving, often including pierced work. They are not joined or assembled afterwards, but carved from one piece of tusk.

To make a puzzle ball, a piece of ivory was taken from the solid area above the pulp cavity in an elephant tusk and cut into a sphere. Several smooth, conical holes were carved into it, almost to the centre. Then followed the intricate work, starting at the centre and working out.

First, the surface of the smallest ball was detached from the surrounding ivory. A tool with a tiny saw attached to the end at an angle was inserted into each hole in turn, gradually cutting the ball loose. Then successive balls were detached in the same way, each one being a little larger than the previous one. The balls were decorated with low relief and intricate piercing as the work progressed. The work was usually done on a lathe to ensure that the balls were truly round, the whole object being repositioned frequently on the lathe in order to carve into it from different angles. Occasionally, damage was done to one of the interior balls, which had to be broken up and carefully removed, piece by piece. However, this resulted in a gap between two of the balls so that the set was no longer completely concentric.

Puzzle balls have probably been made since the seventeenth century, and there have been suggestions that they could have been made two hundred years before that, but they are delicate objects and it is rare today to find an old one that is completely intact. By the latter part of the twentieth century they were being mass-produced and the quality of carving and the designs suffered. Still individually made, but using electrically powered carving tools and lathes, the balls are all of similar size and pattern, often depicting dragons and elephants, and are made for the tourist trade or for export. While the outer ball is still carved in relatively high relief, the inner balls simply have a few pierced holes.

Boxes

One of the most widespread uses of ivory through the ages has been to make or to adorn boxes or containers of some sort. Worldwide, ivory has been chosen for its status implication, its ability to be carved, or the fact that it is a tactile and attractive material.

In past centuries, in places such as the Indian subcontinent where furniture for storage was not the norm, boxes were made and adorned with ivory plaques. In the Arctic, wooden boxes were adorned with little ivory carvings depicting seals, birds, canoes and other items common to everyday life. In Europe wooden tea caddies veneered with tortoiseshell and trimmed with ivory were popular in the nineteenth century, and in the twentieth century plain ivory boxes graced many dressing tables.

Ivory boxes have been used as reliquaries, said to contain a physical remnant of a holy person, or to contain the Holy Sacrament. They have contained unguents and cosmetics, games pieces, calling cards, and a multitude of other things, too.

The methods used to make boxes varied from one country to another. The most extravagant use of ivory was the method adopted in Cordoba in Al Andalus in the Middle Ages, where the boxes were made without joins and were carved from one piece of ivory. The size of the box was limited by the size of the tusk used, and a block was cut and hollowed out. Rectangular boxes were extremely extravagant as they had to be cut entirely from beyond the pulp cavity. Pyxides were slightly less wasteful of the precious material because they could be cut around the cavity, but again the size of the finished object depended entirely upon the size of the tusk used (*see* the diagrams for cutting ivory, Fig. 2.36, in Section 2.7 Working Ivory).

Much more economical use of the material occurred when boxes were made in separate parts and joined together. The sides of a round box could be cut from the area surrounding the pulp cavity, with a slice taken longitudinally from a tusk inserted as a base. This meant that the circular part of the box did not have to be cut from the area beyond the distal end of the pulp cavity, but could be cut from anywhere along the tusk. Square or rectangular boxes could be made up of several slices of ivory that were joined together using ivory plugs, or metal wires or plaques. This method allowed for slightly larger boxes, although the size was still limited by the size of the slice of ivory obtainable from the tusk.

Ivory plaques could be mounted of a wooden core using plugs of ivory, or a form of glue. Depending on the depth of carving, it meant that the plaques did not have to be as thick as if they were themselves the walls of the box. It also meant that several plaques could be mounted alongside each other, so there was no limit on the size of the finished object. Ivory trims for such boxes could be produced in standard lengths and to standard patterns, so although still hand-carved, there was an element of mass production.

In more recent years ivory boxes were made in parts and glued together. They tended to be round, which maximizes the use of the material and minimizes the waste. They were left with little or no adornment in the way of carving, although they often had a simple profile produced on a lathe.

Fig. 2.51 Set of six elephant ivory boxes for the dressing table. Made in England, early twentieth century.

Sundials

Sundials were made in various European cities, notably Nuremberg in Germany and Dieppe in France. They were excellent timekeepers – more accurate than the early pocket watches – but their big disadvantage was that, in order to tell the time, it was necessary for the sun to shine.

Sundials could be made in wood or bone, but by far the majority were made from ivory, with brass or gilt fittings. Most were square or rectangular diptychs – that is, they consisted of two leaves hinged together. A few were oval or round. Sizes varied from very small to a size comfortable in a man's hand.

Inside was a compass, and incised on the ivory were lines calibrated for latitude – either a specific latitude or several – the hours of the day, the phases of the moon and the sun, and the signs of the zodiac. The little instrument also had a gnomon (string or pin) which, when fixed in place, cast a shadow on the dials indicating the time. There were usually two dials, one indicating the direction of the sun's angle, and the other indicating the sun's height in the sky.

▼ **Fig. 2.52** Sundial made by Gabriel Bloud in Dieppe, France, between 1650 and 1680. Partly gilded and painted, with gilt brass fittings.

Some sundials had very complex dials and charts; others were simpler. Those made on commission could be elaborately decorated with patterns or family crests, while others had little decoration at all. The usual colours were black, red, green, blue and a warm brown.

The incised areas were coloured with the usual pigments of mediaeval times, largely made of minerals that were ground to a fine powder and mixed with ground glass and egg white or beeswax to make the colour adhere to the surface (*see* Section 2.7 Working Ivory: Colouring). It is thought that the workers would have been able to buy the colours ready-made in a city of the calibre of Nuremberg.

It is unclear whether the same workshops as prepared the ivory made the metal fittings, but it is quite possible, although the work involved was very different. It would have included magnetizing metal for the compass and, when applicable, gold-plating the brass.

Unless a special commission, the sundials were made for the general public, or at least, anyone who could afford them, and were produced in a variety of standard sizes. They were also exported all over Europe, and even as far as Arabia.

Section 3

Where and When?

The history of ivory worldwide,
from prehistoric times to the present day

Virtually all countries across the world have used teeth and tusks in some form or another, be it as elaborate carvings made from large tusks or strings of animal teeth as personal adornment.

Some countries had their own supplies of ivory (for example, elephants once roamed in China). By the time their herds were depleted the custom of carving the beautiful material was well established, so ivory was imported from elsewhere.

Not all countries had indigenous, ivory-bearing, land animals, and instead they used marine ivory. The very few countries that had no access at all to any teeth that were big enough to be carved still used small teeth as adornment. Teeth and tusks have always held a fascination for people, worldwide.

One or two countries which are now renowned for their beautiful ivory carving – for example, the Philippines – did not develop the custom until relatively late. Ivory carving started there mainly because of demand from foreign settlers. In Japan very little ivory was carved for domestic use, but the craft expanded when the fine carvings caught the eye of foreign buyers.

Carved ivory became a lucrative trade for some countries, while others exported raw ivory. In many cases ivory was traded along with other commodities such as cloth and spices, and the African ivory trade is notorious for being linked to the slave trade.

There was symbolism in much of the ivory used; for example, necklaces of animal teeth might denote that a tribesman was a good hunter. They could also have a spiritual representation. The use of large, ornate ivory carvings have generally denoted status, wealth and power, and these were originally owned almost exclusively by the ruling families or courts, or by religious establishments – notably Christian.

The most avid users of ivory were, in times past, the Europeans. Today most ivory carving is done in the Far East and is a mixture of legal and illegal work.

3.1

Raw Ivory Trade Routes

Ivory has been traded for thousands of years. It was seldom the only item in a cargo, but was more commonly sent together with other luxury goods such as silks and spices, minerals and other desirables of the opening world. It was also sent together with slaves, but even then it was often considered the most important part of a cargo because it was deemed to be so valuable. When once the beauty of ivory had been discovered, people had to possess it, and when their own supplies ran out they found ways of obtaining more – by looting, by demanding tribute, by barter, and by trade.

The colonization of far-flung countries was based largely on the wish to trade. The routes taken were fought over and jealously guarded – and they depended largely on the weather. Trade was not simple.

In the case of ivory, trade was always connected to power and often to religion, as settlers in new countries desired to have their religious iconography made from the precious material.

Ivory was carried over land by horse, camel, and humans. On sea it was carried on board canoes, junks, dhows, triremes and quinquiremes, sailing ships and steam ships. Later it went by road, by rail, by cargo ships and by air. It was sent by every conceivable form of transport to get it to the people who wanted it.

When their own supplies from local animals were insufficient, the Ancient Egyptians ventured up the Nile in search of more ivory, going beyond the borders of their neighbours, Nubia, and possibly as far as Ethiopia.

In Classical times ivory was traded from Ethiopia to Greece and the Roman Empire. It was carried up through the Red Sea and on via Egypt to the various Mediterranean destinations, and old writings imply that there was a lively trade

▲ **Fig. 3.1** 'The Ivory House' at St Katherine's Docks, London, built in 1858 to accommodate the huge imports of ivory arriving at the docks.

up and down the coast of East Africa. Cargoes included ivory and tortoiseshell, rhino horn, and palm oil.

An alternative route was across the Sahara Desert by camel train. The caravans carried a variety of goods to ports on the north coast of Africa, notably Alexandria, for onward transport to other ports around the Mediterranean.

The trade from East Africa is often described as having been controlled by the Arabs, but this is a simplification of the facts.

By the eighth and ninth centuries, the East African coastal area was populated by the Bantu, predecessors of the people known today as the Swahili. Arabs who had fled the caliphate and settled in East Africa had intermarried with the local population, and with time the Swahili society evolved. Some of them were excellent traders and they had, for the most part, good relations with the Arabs and the Indians who wished to trade with them.

By the time the Roman Empire collapsed, demand for ivory for India and China had increased, and tusks were being sent across the Indian Ocean via ports on the Arabian Peninsula to ports in India, and on to China. The trade with central Asian areas was already established, using ports in Arabia or Basra on the Persian Gulf, and some ivory was still sent up the Red Sea to be transported to Alexandria or to the new Islamic Egyptian capital of Fostat on the Nile (now part of Old Cairo) for onward transit to ports such as Byzantium and Venice. The trade routes effectively extended from China to Europe by routes across the Indian Ocean and the Mediterranean Sea.

Dhows crossed the Indian Ocean with the monsoons. From November to February the Arab and Indian traders could sail to East Africa, but from May to September the winds turned and they were able to sail back home. This gave them April in which to trade. If by chance they missed the May sailing, they preferred to stay until September before returning home as the winds from June to August were too fierce. Thus the trading vessels arrived annually, allowing the Swahili, with whom they traded, to build up stocks in the interim.

By the thirteenth century ivory was pouring into Europe. It came from East Africa and on the trans-Saharan route from West Africa. Al Andalus (the mediaeval Islamic Iberia, comprising most of what is today Spain and Portugal), had gradually been retaken from Muslim control and had reverted to Christianity. Trade could again pass freely through Spain on its way north. A passage had opened through the Straits of Gibraltar, allowing further shipments to travel from Majorca, which was a trading hub in the Mediterranean, to ports on the Atlantic, for onward transit to cities such as Paris. In England, ships could sail up the Thames to London to unload their cargoes.

Not content with being at the end of the trade routes, and intent on expanding their empires, the Europeans were keen to get in on the action. At the end of the fifteenth century and beginning of the sixteenth, the Portuguese arrived in East Africa, with the aim of taking over all the lucrative trade routes in the Indian Ocean, and colonizing the area. They were in fact more interested in gold and spices than in ivory, but nonetheless they included ivory in their cargoes. Somewhat uneasy alliances were made, but despite many battles the Portuguese never managed to dominate the trade routes, only succeeding in disrupting the Muslim control of trade by seizing coastal ports. The various trading ports on the east coast and its islands included such places as Mombassa, Mozambique, Zanzibar and Kilwa.

The Dutch, British and French were also interested in the Indian Ocean trade routes, which were no longer simple journeys from home to a foreign country and back, but had become carrying routes – that is to say, cargoes were transported to various ports, the goods being sold wherever there was a demand. The Dutch had colonies in South-East Asia, and trade was now going in all directions, with ivory being shipped to even more destinations to satisfy the demands of the colonists who had brought with them their requirements for religious carvings in ivory. London became one of the main ports for the import of raw tusks and, in the early seventeenth century, about 5 tonnes was arriving there a year. (Three hundred years later London was still the main importer of ivory, but the amount arriving is estimated to have increased a hundredfold.)

By the late seventeenth century the Omanis had also started colonizing, and were taking over Portuguese colonies in East Africa. They had been involved in the ivory trade for a long time as they had vital ports in their own country on the Arabian Peninsula, through which the trading vessels passed, but they were also keen to have a foothold in East Africa themselves. With time they developed a large settlement in Zanzibar which became the most important point for export up until the last quarter of the twentieth century.

Meanwhile, the Indians were also sailing their own ships and ceased to rely on the Arabs to act as middlemen for their trade. The Indian trade had been largely under Arab control, and there were Arab settlements on the Indian west coast – for example, at Cambay (now Khambhat) in Gujarat. With the other countries busy fighting amongst themselves, the Indians quietly took over domination of the ivory trade. Their position was strengthened by the British presence in India, and furthermore they gained control of the financial position of the Omanis, whose main income was from the taxation of ships using their ports. The Indians continued to dominate the ivory export from East Africa for the next two hundred years.

All the facts stated so far apply to elephant ivory. It is worth noting that hippo ivory was also being traded, and following the same routes. It is not known how much hippo ivory was exported, but it is certain that the amounts in no way compared with the amounts of elephant ivory, which was far more highly prized. Although there are records in some European markets of ivory purchases which differentiate between the two species, there was apparently seldom any differentiation made at the point of export, so little is known about the trade of hippo ivory within and from Africa.

When elephants became scarce on the African coastal regions, expeditions set forth to the interior to find more. They went partly on foot and partly by canoe. A canoe could, for example, travel 400 km up the Zambezi River.

The elephant hunters came from various societies, and they traded and bartered the ivory with the Swahili, who in turn traded with the foreigners. One such society was the Yao people. They were agriculturalists who also hunted. They lived inland but came to the coast to trade. Originally they had only been interested in obtaining good-quality salt, but they progressed to trading for metal implements, cloth, tobacco, and other things to make their daily lives easier and more pleasurable. Finally, they progressed to trading for guns.

By the early eighteenth century firearms were beginning to appear as trade items. These were at first viewed with suspicion by the African peoples, not least because the guns sent to Africa were often defective and could easily explode on firing. It was not until the latter half of the nineteenth century that reliable rifles were being imported. Firepower changed the face of both the ivory and the slave trade, and hunting elephants was no longer the dangerous occupation that it had previously been. Furthermore, greater numbers of elephants could be killed.

After a while the foreign traders also ventured inland, developing settlements and – at least in the case of the merchant class – enjoying a good life. The ruling merchants lived in stone houses, with plenty of slaves to look after them. As the elephants were hunted almost to extinction in nearby areas, hunters had to be employed to travel further and further inland in order to find the ivory, so the settlements acted as trading posts.

The increased trade brought increased wealth to some Africans, although it was not monetary but in the form of goods such as copper and brass, and it was mostly the rulers who benefited. Some of them also built up huge stockpiles of ivory tusks as an additional form of wealth.

Elephant hunts were a skilled occupation and were risky, but they were well organized. In fact, in some societies the hunters were organized into guilds. The hunt needed the approval of the chief before setting off, and he in turn had to get the approval of the ancestors – a task which was accomplished with a ceremony. The hunters usually wore amulets to protect themselves, possibly wrapping more amulets around their spears (and later, guns), and rubbing potions into their skin. The hunts could last for some months, and there were festivities when the hunters returned home laden with elephant tusks. The elephants had become fair game and the myth and magic of the great beasts, which had been so strong in Africa in times past, became of little value. By the time of colonization it was all about money.

There are many stories of men being forced to carry ivory to the coast, and then being sold as slaves when they arrived there. Although this surely happened, it appears not necessarily to have been the norm, as some of the porters were the merchants' own slaves, and in other instances ivory-carrying was a respected job for which the porters were paid. Also, acquired trade goods needed to be carried back inland. However, trade in slaves was being conducted at the same time as the ivory trade, and usually under the orders of the same merchants. The two trades converged and both slaves and ivory were sold and transported in the same vessels.

For a while the slave trade overtook the ivory trade in importance. It had been going on for centuries, but on a much smaller scale. As the demand for workers grew in American and Europe, so too did the trade in slaves, and their value increased. The ivory trade was then secondary, but it continued as it was still lucrative. When slavery was banned in the middle of the nineteenth century by the British, who were trading on the West African coast, many ships took the perilous journey around the Cape of Good Hope to fetch slaves for the Americas and other destinations from the east coast, where the trade was still permitted by other nations. When slavery was banned altogether, the ivory trade continued on both east and west coasts.

Possibly as early as the fifth century AD – or maybe even earlier – much of the ivory and many of the slaves that had been arriving in Europe from West Africa had come via the arduous and dangerous trans-Saharan routes. From various trade regions in West and Central Africa, slaves and ivory had been collected at trading hubs such as Gao and Timbuktu, from whence they travelled north along the desert caravan routes to ports on the north coast for their onward journeys to other Mediterranean ports. But from the sixteenth century, it became common to navigate the Atlantic in sailing vessels. The main ivory trading ports were in what are known today as Côte d'Ivoire and Ghana. The trade was started by the Portuguese, and the combined trade of ivory and slaves continued – the so-called 'white gold and black gold' trade – to the markets of Europe and the Americas. This could be done without competition or interference from the traders on the east coast, where the Portuguese had faced such stiff competition, and was proving immensely lucrative.

The west coast trade expanded when the Dutch and English arrived, opening up ports further down the coast. Various routes emerged, including the 'trade triangle', which involved ships sailing from the African west coast to the east coast cities of North America, on to Europe, and back to Africa. Their cargos were slaves and other goods, including ivory, which was now being brought from the hinterland to several ports along the coast.

Africa was not, of course, the only region that could supply elephant ivory. India also traded ivory, though on a much smaller scale. In about AD 200 China's elephant population was finally extinct, and the Chinese began importing ivory from nearby countries, those that are today called Myanmar, Thailand, Vietnam and Cambodia. Trade with these countries continued unabated, but as demand for the precious material increased, as previously mentioned, China began importing ivory from Africa as well, via India.

India had its own tradition of carving ivory, and as ivory began to arrive there from the east coast of Africa on its way further east, the Indian craftsmen began to favour the African material. Some say that they found it easier and more satisfying to work, but the truth is probably that the tusks were much larger than those from local elephants. India had become a trade hub and was both importing and exporting elephant tusks. The main ports were Kachchh and Gujarat. It is probable that Asian and African tusks were mixed together. As it is impossible to differentiate by sight between the two when carved, today it is not possible to tell without specialized testing whether many of the old carvings are made from Asian or African ivory.

Plenty of puzzles still remain concerning the movement of ivory for trade. What was a cache of hippo teeth, ready to be packed and sent off, doing buried in the ground in Gao in Mali? Where was it going? And where was it from? There is a theory that it may have been destined to be used as inlay. Hippo ivory is very homogenous in appearance and lacking in clear structural patterns, but more importantly it retains its white colour and does not have elephant ivory's tendency to darken, which would have been an advantage for inlay. Another cache was found, this time of elephant tusks, at a place called Wad ben Naga, by the Nile in the Sudan. It was most likely headed north and intended for trade.

The question of trade to and via Al Andalus has also been a puzzle and the subject of various theories. It was thought that all ivory reaching the region came from ports on the east coast of Africa, but today there is convincing argument for the fact that it took the trans-Saharan route previously mentioned. It is not believed that any came from India. Also, some of the ivory reaching Al Andalus was not trade, but was sent as tribute. For example, old writings state that a Berber prince sent to the Spanish Caliph a gift of many animals, animal skins, and 'eight thousand pounds of the purest ivory'.

A long-held belief is that the decline in available elephant ivory in northern Europe in mediaeval times was a result of the Islamic conquests, notably of Al Andalus. Yet other trade goods continued to get through, so perhaps for a time walrus ivory was the preferred material because it was less expensive. During that time elephant ivory did continue to be used, albeit in much smaller quantities, and the general opinion is that it came from stockpiles of the material, or from old ivory items that were reworked. But there is a possibility that some of it was, in fact, fresh ivory, recently arrived from Africa. If there was indeed an embargo on ivory passing through an Islamic country, it is more likely to have been an edict made by the Pope in Rome, which forbade the trade. There is evidence of an instance of ivory arriving in Venice and being confiscated because it had passed through the hands of Muslims.

Trade hubs and destinations depended on wealth, and also brought wealth. Ivory was always a luxury and, as mentioned, was seldom the only item in a cargo. It was sometimes carried together with alum, the mineral mined in parts of Africa, which was the fixative used in dyeing fabric. Cotton and silks made up more of the cargo. Thus in some instances the ivory trade was connected to the textile trade, and it arrived at centres where there was a textile industry, for example, Rheims. Carving centres developed in these areas, supplying the churches and cathedrals around which the towns had been built.

Ivory was, of course, also carried with spices, the all-important cargo that was so sought after as spices could make bland foods taste better and, with their antimicrobial properties, could help to preserve it, or at least hide the taste of rotten food.

The trade in walrus and narwhal ivory is not as old as that of elephant ivory, largely because it was found in inhospitably cold places where the inhabitants were very insular. Thus, for example, although the Inuit in Greenland had used ivory for some four thousand years for their own (largely utilitarian) purposes, it was not until the arrival of the Norsemen in 985 that the tusks became a feature of export. Instead of living as nomads, following their prey, the Norsemen settled in south-west Greenland. They were not able to exist in the harsh conditions on what was available locally, so trade was imperative. They had trade connections with the North Atlantic region of Europe, and imported items such as iron and wood. They exported seal fur and walrus skin, which made excellent, strong rope. However, their most valuable export during the first centuries of settlement was walrus tusks. These were much in demand further south when elephant ivory was, for whatever reason, apparently no longer as abundant as it had been.

Norsemen not only hunted the animals themselves, they also traded with the Inuit walrus hunters. In return for furs and tusks the Inuit received iron tools. Walrus ivory was, at that time, already being sent to western Europe from Russia and northern Norway, along well-established trade routes.

Narwhal ivory was also a sought-after trade item. As mentioned in Section1.6 Narwhal, the Europeans thought that the animal's tusk came from a unicorn. This rare and magic 'horn' could purportedly detect poison, cast spells, and had very strong talismanic properties, and so was worth its weight in gold. The people of Greenland did not disillusion their trading partners, and made the most of the myth until it was dispelled in the mid-1660s.

Trade in raw walrus tusks virtually ceased when elephant ivory again became plentiful in Europe. It is thought that by the middle of the fifteenth century the Norsemen had gone completely from the whole of Greenland, having given up the struggle of living an agrarian life in a climate that was getting colder (owing to a change in climate at the time). Already a hundred years before, a Norwegian priest, Ivar Bardarson, who had visited southern Greenland in 1343, wrote that he found 'horses, goats, and sheep, all wild, and no people, either Christian or heathen'. There was little trade for the next few centuries, and the Greenland Inuit were largely ignored until the Danish king took an interest in the eighteenth century, and Greenland became a Danish colony from 1721. However, the trade in walrus and narwhal ivory from Greenland was never again of much importance.

In the seventeenth century the Canadian Inuit traded a few goods with the British and French settlers, although they were not averse to occasionally raiding the settlers' camps to get what they wanted. By the eighteenth century they were trading for firearms, with which they could hunt many more and various animals in order to trade, although by this time the most popular trade goods were no longer tusks, but beaver pelts.

Records state that the Inuit of the Bering Straits traded walrus ivory with the Chinese in exchange for iron utensils. Why walrus ivory came to be used in Hawaii may be more of a puzzle, but it was used for ceremonial pendants, hung on cords made of human hair, or made into bracelets (*see* Section 3.10 Australasia and South-East Asia). Sperm whale ivory had originally been used to make these objects, but the people of Hawaii were surely delighted to be able to obtain the much larger walrus tusks when they were offered to them in trade by the visiting whaling ships, even though the spiritual connotations were lost by using a material that was not local.

Marine ivories were a by-product of the trade of the whaling ships. Various countries hunted whales near their coasts, but pelagic whaling from ships started early in the seventeenth century, so by the time the Japanese of the Meiji period were carving small ivory objects (for example, *netsuke* and *okimono*) ships were sailing and trading far and wide and the fact that some of these objects were carved in walrus ivory is less surprising. The material was, on occasion, used by choice to give extra character to the carving, rather than out of necessity.

Today there are very strict laws worldwide regarding the trade of any form of ivory from extant species of animals, including that which is derived from the legal killing of the subsistence quota of animals by certain indigenous groups such as the Inuit. However, the trade in ivory continues, and the routes taken today are ever-changing and secretive. As in earlier days, ivory is seldom the only item in a cargo, but now that is not to lessen the cost of transport, but in order to conceal ivory's presence in the cargo (*see* Section 1.1 Conservation).

The trade in mammoth ivory is, of course, legal because the mammoths are already extinct. The workable material comes from areas covered by permafrost, as the material from warmer areas is too degraded to be of any use. Alaska and Canada use a little of the material and also export it. Siberia is the main supplier, also keeping a little for domestic use but exporting about 90 per cent to Asia.

3.2

Prehistoric Times

Art was not unknown to the people who lived many thousands of years ago. They decorated the walls of caves with pictures of life around them, so their subjects were mostly people and animals. In Europe, Russia and America there were abundant pictures of mammoths which, along with bison and horses, were the favourites to illustrate.

In most cases the walls were engraved or painted, and in a few places the pictures were carved in relief.

Much of our supposed knowledge of prehistory is conjecture. It is not possible to know for certain how things were accomplished, or what their purpose was, but evidence that remains can give some strong indications. For instance, it is believed that as far back as the Neanderthals, tools were made out of mammoth bone. Neanderthals also decorated mammoth molars with pigment, although the significance of this is unknown.

Mammoth bones and tusks were certainly used as building materials. The huge skulls and the larger bones formed the basic structure, and the smaller

Fig. 3.2 The Swimming Reindeer. The tip of a mammoth tusk carved as two reindeer, one behind the other, swimming. They were found in the Midi-Pyrénées region of France in 1866, and are at least 13,000 years old.

bones and tusks supported the walls and roof. Presumably the whole construction was covered in hides. Remains of circular and oval huts made in this way and dating back 18,000 years have been found in Eastern Europe. The hides have rotted away, and the rest of the dwellings have collapsed in on themselves, leaving just a large pile of bones and tusks. Over twenty skulls could be used on one hut, so the skeletons must have been plentiful. Not all the materials were derived from freshly killed animals: in fact, carbon 14 dating has shown that, in one hut alone, the bones differ in age by 10,000 years.

Music is thought to have played a part in the life of early man, as musical instruments have been found where they were living. Some were carved from bone, but some were also made from mammoth tusks. The earliest one known was made of ivory and found in a cave in south-west Germany. It is 31,000 years old.

Basic furniture and articles for human adornment have also been found, made from bone, antler, ivory or stone, yet it is not possible to know for certain how these items were carved with the primitive tools available at the time, which consisted basically of sharpened pieces of flint.

When a mammoth had died of natural causes the tusks would, with time, become loose in the jaw. If it was freshly killed the tusks would be firmly fixed, but apparently people were able to remove them from the skull using large flints as axes.

Cutting the tusks up must have presented another problem. Mammoth tusks are big, extremely solid, and very tough. It is not possible to knock off a small piece simply by hitting it, and a very heavy blow with a large rock would only succeed in splintering the tusk. Various theories have been put forward as to how the tusk was cut. The first is that it was chiselled around the circumference with a sharp flint, until just a narrow piece remained at the centre, at which stage the section could be knocked off. A second theory is that, having chiselled a groove, a flat stone was used as a saw, using sand and water for abrasion. Another theory is that grooves were chiselled in the tusk, and small slivers of wood inserted. When these were soaked with water they swelled and caused the tusk to crack. As ivory is hygroscopic and the tusk would also have been affected by the water, this method seems a little less likely. In all probability a mixture of methods was used and there was no definite system for cutting the tusks.

Cutting smaller pieces from the section of tusk was probably done by tapping it repeatedly with a sharp point. Striking it hard would again have caused it to splinter, but, done gently, it is possible to flake off pieces, although it is unlikely that the exact size and shape of the piece could be predetermined.

Having removed a chunk of ivory of approximately the required size from the tusk, sharp stones were used to carve or engrave it, and anything abrasive could be used to polish it. A final shine was probably given with a soft piece of hide.

Stone Age man made many utilitarian items out of ivory. Among those that have survived are tools with a flat, sharp end to be used as a chisel, or with a sharp point, possibly for incising. Archaeological digs have also exposed barbed harpoon points, needles, and scrapers to clean hides.

Graves dating back 30,000 years have been found to contain skeletons decked out in thousands of ivory beads and bangles. Half-finished beads have been found, which appear to have been made by cutting small rods of ivory, whittling them down at intervals, snapping the sections off, and finishing each bead individually.

The items that perhaps catch the imagination more than any others are the small figurative ones made from bone, antler, or mammoth ivory, and carved in the round. Usually they depict the animals which would have been known to the carver, in the same way as the cave paintings. Although not very detailed, the subjects are instantly recognizable.

Some carvings depict figures that are half-human and half-animal, such as the famous 'Lion Man' found in pieces in the Stadel Cave in Germany in 1939. At the time the importance of the find was not known as, because of the advent of war, the ivory fragments were not properly examined. Having been left for thirty years, when re-examined it was discovered that about two hundred pieces actually fitted together in the shape of a standing human figure with a lion-like animal head.

A popular subject for carving was the female figure. It was usually depicted totally nude and with very exaggerated, sometimes pregnant features. The purpose of these so-called 'Venus' figures is unknown, but a plausible suggestion is that primitive man considered it magical when a woman became fat and then produced a living being. Such magic would have been worthy of recognition. Venus figures were also carved in stone, and have been found all over Europe. The Venus of Hohle Fels was discovered in 2008 near Schelklingen in southern Germany. It is thought to be between 35,000 and 40,000 years old and is the oldest representation of a human figure ever found.

The flints used to carve such figures would have had to be sharpened regularly as they would have quickly become blunt, and the ivory was most likely wetted before carving to make it slightly softer. Accounting for the slow pace of work with such tools, and working only in daylight, a reasonable estimate of time taken to complete a small, portable carving would have been a minimum of two months of fairly constant work. If carving was simply a pastime, the creation would have taken a lot longer.

3.3

Palaeo-Eskimos, Inuit and Norsemen

In the far north, within the Arctic Circle, the ivory-bearing animals are narwhals and sperm whales, orcas, and the most important source of ivory – walruses.

The Palaeo-Eskimo people came mostly from Siberia. They travelled east, some settling in what are now Alaska and Canada, and others continuing on to Greenland, arriving there in about 2500 BC.

The Inuit culture arose in the Bering Strait region in about AD 1000. All the various Inuit groups had similar customs and used different dialects of the same language, as is still the case today. In 1970 the name 'Eskimo' was dropped by the Canadian and Greenland populations in favour of the name 'Inuit'.

The exact origins of the various Eskimo cultures are uncertain. The first settlers are referred to as Palaeo-Eskimos, a term that applies to all the early cultures, culminating in the Dorset Culture, which arose in the Canadian Arctic in about 500 BC. While earlier Palaeo-Eskimos had lived more inland and hunted caribou and musk oxen with bows and arrows, the people of the Dorset Culture lived primarily along the coasts of northern Canada and Greenland and hunted seals and walruses with harpoons. And they used ivory.

The Dorset Culture was named after Cape Dorset in Nunavut, Canada. The people lived until about AD 1300, and then seemingly disappeared, although no one really knows why. It is possible that a change in climate had some effect, or also that they were driven away by new settlers or totally absorbed into the settlers' culture.

Dorset Culture ivory carvings included animals and human figures, and sometimes a mixture of both. They were engraved with a pattern resembling the figure's skeleton. Utilitarian objects were also made of ivory, and included such items as harpoon points, arrowheads, and snow knives.

The people of the Okvik and Ipuitak cultures lived in north Alaska and north-east Siberia respectively. The earliest Eskimo civilization of which material evidence has been found is from the Okvik people, who lived from about 450 BC to AD 900. Archaeological excavations have unearthed middens or rubbish tips containing many carved ivory artefacts, among them lots of small figures that have human faces but no arms or legs. Their purpose is unknown.

By contrast the archaeological finds from the Palaeo-Eskimo Ipuitak people were in burial chambers, not middens. The Ipuitak carved some remarkable openwork figures, and also apparently adorned the dead with pieces of ivory. The Ipuitak lived until about AD 700. Their origins are unclear but they probably were North American.

In about AD 1000 a new wave of settlers arrived via the same route as taken earlier by the Palaeo-Eskimos. They became known as the Neo-Eskimos, and were the so-called 'Thule Culture', from which today's Inuit are directly descended. The Dorset and Thule cultures overlapped, and the people may have lived alongside one another for a while in the eastern Arctic – that is, Canada and the north of Greenland – but there is nothing in writing to confirm this. The legends and the history are purely anecdotal, handed down over subsequent centuries. What is known about their cultures comes largely from the burial sites, and middens unearthed in archaeological excavations. Both tell us something about how and where the people lived.

It is known that the Inuit were animists, and this greatly affected their use of ivory. Animists believe that everything has an energy, be it a human, an animal, a rock, or a tree, and that therefore everything must be respected, and a balance upheld. It was accepted that in order to survive, the Eskimos had to kill animals. It was assumed that the animals understood this, but they were killed with the greatest respect, and because of this would have been pleased to give up their lives in order to keep the balance. The respect was shown partly by the tools and weapons used to kill the prey. For example, the handle of a knife might be carved with an image of the intended prey, such as a seal. This was not to please the owner of the knife, but to honour the soul of the seal, as the spirit of the animal killed would return in another animal of the same species. If the seal was angered and not respected its spirit would not return to repopulate the waters.

There was a strong belief in spirits. Anyone could call upon spirits to help them in times of need, but the spiritual leader, a shaman, possessed strong magic and was the medium through which the people and the spirits could communicate. He could intervene on behalf of the people if they failed, as it were, to keep their side of the bargain. Furthermore, a shaman could take on

Fig. 3.3 Sun visor: collected in 1900–10 from Ammassalik, Tasiilak, Greenland.

the form of an animal, especially of a polar bear, which was the most respected animal in the Arctic.

Inuit were superstitious, and felt that danger lurked everywhere. The use of amulets was therefore widespread, and not only were they worn, they were also placed in kayaks, under beds, and so forth to ward off evil and keep the people safe. Leather belts to go around the waist were hung with a variety of teeth including reindeer and polar bear, also bones, and birds' skulls.

These traditions are reflected in both the Palaeo-Eskimo and the Inuit ivories that have been uncovered. There are small carvings in the shapes of the animals which they hunted, probably made in honour of those animals. There are boxes for storing the tools for hunting, decorated with tiny ivory carvings of animals, humans and kayaks, again honouring the animals against which the tools would be used. And there are small carvings of human figures with some animal characteristics, thought to be aids for the shaman when he was changing into another form. It is likely that every carving and its decoration had a symbolic or talismanic meaning.

Walrus ivory was the ivory of choice for working because of its size. Some narwhal ivory was used, but its shape precluded many uses. Occasionally, a very straight tusk could be used as part of a harpoon, though they were seldom straight enough for that, so generally only small items could be carved from them. Sperm whales and orca were plentiful, but too large to catch from kayaks, so their teeth were little used.

In spite of their size, whole walrus tusks were apparently too valuable to be used as building materials. On the other hand, narwhal tusks were used as tent poles. Nothing goes to waste in an environment such as the Arctic, and all parts of an animal were put to use. The Inuit were reliant on sleds to move about, and in order to make them run smoothly pieces of polished ivory were attached to the runners. All the Arctic ivories made excellent harpoon points as they could be carved with barbs and sharpened to a point. Some of the Palaeo-Eskimos devised very effective ivory harpoon points in three sections, held together with fine rope made of leather. When the point had entered the animal's body it could not be pulled out or rubbed off, so it was impossible for the animal to escape.

Being large, a piece of walrus tusk had many uses. There are examples of sun goggles that date back to AD 500, made by the Neo-Eskimo Punuk people who lived in Alaska around the Bering Sea until about 1200. The goggles had small slits for the eyes, and protected them from the glare of the sun reflecting off the snow.

Eskimo and Inuit children's toys were made of ivory. Examples are miniature sleds drawn by a pack of miniature huskies, and babies' rattles consisting of a few polar bear teeth strung on a piece of leather. Such a rattle surely had the added value of being regarded as talismanic and so would keep the baby safe.

There are very few examples of polar bear teeth being carved, although they are large enough. When used, the teeth simply had a hole pierced in order to string them, but bore no decoration, probably because the polar bear was considered too powerful and was too much respected – indeed, some of the Palaeo-Eskimo cultures were very bear-orientated, believing them to be on a par with humans (*see* Fig. 1.59, polar bear tooth pendant.)

The early Palaeo-Eskimos had no iron tools with which to work ivory and had to make do with sharp stones. This would have been a very long and laborious job, probably undertaken in winter when there was time for such work, although the lack of daylight hours cannot have made the task any easier. Excavations from the Neo-Eskimo Punuk sites showed that iron was imported and used for tools, but where other civilizations used wooden handles, the people of the Arctic tended to use ivory, in many cases decorated with incised pictures or patterns.

Hunting was not a winter occupation. The meat from the animals killed during the rest of the year was stored in the ice, and their tusks stockpiled until they were needed either for the people's own use or for the arrival of the traders in the spring.

Prior to their region being taken over by the Russians in the sixteenth and seventeenth centuries, the Siberian Inuit traded with their neighbours in Russia.

The Canadian Inuit traded with their neighbours to the south. The story was different for Greenland Inuit because Greenland is an island and access was only possible over the polar ice in winter, or by foreign ships when the ice around the more southerly parts of the country melted in spring. Following the arrival of the Norsemen at the end of the tenth century, the inhabitants of Greenland traded walrus and narwhal ivory with foreign countries.

The Norsemen brought with them their different social system of land ownership and Christianity, and they built settlements where they farmed. Their survival relied heavily on trade with Scandinavia because they lived in one place and did not follow their prey, living a day-to-day existence as did the Inuit. Church taxes were introduced, which were mostly paid in walrus ivory, and the main exports were walrus tusks and rope made from walrus hide. These were in great demand in mediaeval times in Europe and fetched high prices, because elephant ivory was scarce at the time and the rope was incredibly strong. The Norsemen traded with the Inuit (whom they called 'Skrælinger' or 'puny creatures') to get the ivory and hide, and then traded them on with the Europeans. By the thirteenth century the trade had dwindled as the Europeans were again importing plentiful supplies of elephant ivory from Africa – which they usually preferred to walrus ivory – and virtually stopped buying the latter. The value of walrus ivory fell – a fact which may have contributed to the Norse settlers abandoning Greenland, having lived there for some four to five hundred years.

The Inuit continued with their nomadic life and continued to use ivory, but regular trade was not resumed until the Danes arrived at the beginning of the eighteenth century. By that time walrus ivory was no longer of much interest, and nor was narwhal ivory because, by then, the rest of the world had realized that narwhal tusk was not unicorn horn, but merely a curiosity made of ivory (*see also* Section 1.6 Narwhal *and* Section 3.1 Raw Ivory Trade Routes).

There was, though, a somewhat irregular form of trade from visiting whaling ships. In the nineteenth century whaling ships were obliged to travel far and wide to find their prey. The sailors on board were keen to purchase walrus tusks to carve and scrim and the Inuit were keen to buy iron tools. The whalers also wished to purchase finished carvings, so the Inuit were quick to supply them, in the form of figures or boards for games such as cribbage. Rather than carve them to European styles, they were made in typical Inuit style, with small animals adorning the surface. However, the spiritual connection between the carver and the animal whose tusk he was carving, and the symbolism of the item, were lost.

Fig. 3.4 Cribbage board, carved from a whole walrus tusk. Probably Alaskan, late nineteenth century.

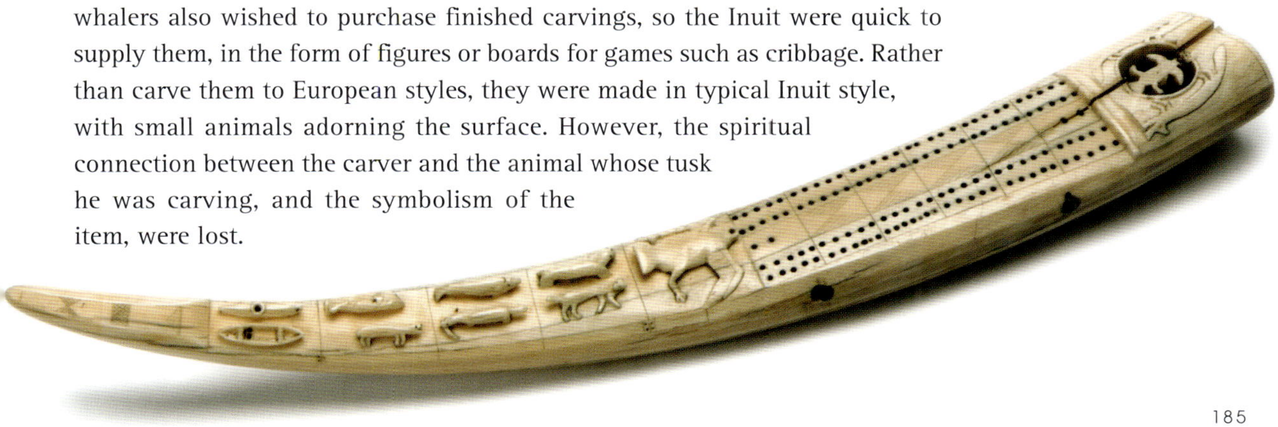

In Greenland the Inuit had one form of ivory carving that was unique: the *tupilak*. These little carvings are usually half-man, half-beast, and were deemed to be imbued with strong magic. Far from being talismanic, these little carvings were intended to work evil magic on specific people.

A *tupilak* was made in secret by a person with strong magical powers. The carving was buried for a period of time along with something belonging to the proposed victim (for example, some hair, plus various potions). It was bound in moss and had spells cast upon it. When it was delivered to the person who had commissioned it and intended harm to his enemy, it was kept well hidden and secret. The process was, however, something of a gamble, because the magic spirits of the little carvings were capable of turning around and bringing evil to their new owner instead of to his proposed victim.

Originally the *tupilaks* were made of bones, sinews, fur and teeth from various dead animals bound together, but later they were carved in sperm whale ivory or reindeer antler, possibly more for the curious European visitors than for magic. Today old ivory *tupilaks* are collectors' items, but there are many much newer copies and imitations around. These are usually produced in the Far East, and are digital copies of old, genuine *tupilaks*. Many Inuit still retain great respect for the spirit world, and an Inuit would be unlikely to wish to display a *tupilak*. To the rest of the world they are regarded as valuable curiosities and are proudly exhibited.

The Aleut people from the Aleutian Islands – a chain of islands stretching from the Alaskan Peninsula to the Kamchatka Peninsula in the northern Pacific – are culturally related to the Inuit. They share some of the customs of the Inuit, but speak a different language. Their use of ivory was not huge, but they are renowned for their hunting hats made of thin sheets of bent wood, which are adorned with sheets of incised ivory and sea lion whiskers.

▲ **Fig. 3.5** *Tupilak* carved from sperm whale tooth. Greenland, early twentieth century. Height 10 cm.

► **Fig. 3.6** *Tupilak* carved from narwhal tusk. Note the spiral morphology of the tusk, visible on the creature's back. Greenland, early twentieth century.

3.4

Europe

The European love affair with ivory began many centuries ago, and it could be argued that it is, at least in part, thanks to the Europeans' love of the material and their colonization of many countries in the world, that ivory became so popular the world over.

Whereas in some countries or societies styles of carving remained the same for many centuries, dictated by tradition, in Europe the carving styles were constantly changing and developing.

Classical Antiquity

Large amounts of ivory were used in Ancient Greece. Magnificent statues were made for the temples and treasuries using sheets of carved ivory and beaten gold laid over a wooden base. The statues, which were numerous and many of which were much larger than life-size, depicted the gods. The most famous examples are the statues of Athena in Athens and Zeus in Olympia. The latter, which is said to have stood about twelve metres tall in spite of the fact that it was a seated figure, was considered to be one of the Seven Wonders of the Ancient World.

Ivory represented skin, and gold the clothes and the hair, in a combination called 'chryselephantine'. It was said that in times of need, the gold could be stripped off the statues and melted down to be used as currency, and replaced when times were good again. The statues are long gone, but a few fragments have been found in archaeological digs. Some were found in Delphi in 1939, but it is thought that the fragments found at Olympia, together with tools for working them, originated from the workshop of Phidias. He was the most famous of the sculptors at the time, and is credited with having made the statue of Zeus. He also made sculptures in bronze.

Another use of the expensive material was as carved panels for furniture, especially couches, chests, and low tables. Again, only fragments have been found. Among the popular motifs carved were mythological scenes. Ivory was also used to veneer chariots, and a little ivory jewellery was made, sometimes incorporating amber.

Ivory was held in extremely high esteem, and was a status symbol. It seems that, though expensive, elephant ivory was plentiful. It was imported from Ethiopia and Libya, and possibly also from further afield. A little hippo ivory is also mentioned in old texts, but the bulk of the material used was from elephants.

During the first two centuries AD there was some controversy regarding elephant tusks, and whether they should be classified as teeth or horns. Many respected writers had toyed with this problem, including Pliny the Elder, who went to some length to compare the dentition of elephants with other animals. He was firmly of the belief that tusks were teeth, even though he stated that the elephant has fourteen teeth apart from the two that protrude, when in fact they have four with another four possibly beginning to grow through (*see* Section 1.3 Elephants).

On the other hand, Oppian of Corycus was in no doubt that the tusks were horns. This was based on various observations, but the most convincing in his mind was that 'growths from the upper jaws of wild beasts as are horny spring upward. If they incline downward, they are certainly teeth'. This would have meant that walruses had teeth, but elephants had horns.

The African ivory being imported in Roman times was brought into the port of Ostia, the harbour city of Rome. Various archaeological finds also indicate that there was good trade between Rome and the Indian subcontinent, and both raw tusks and finished ivory items arrived in the Roman Empire.

It has been suggested that the Romans were just as happy to use bone or antler as ivory, and that they treated all three materials the same. The truth is more likely to be that bone and antler were plentiful and local, while ivory was expensive, scarce, and had to be imported, so that when ivory was not available they made do with bone and antler instead. This conclusion is underlined by the description of a Roman general returning victorious from Syria in 189 BC. In his triumphal procession through Rome was carried silver, gold, and 1,231 elephant tusks – all of which he had brought back from his campaign.

In AD 77, in his *Natural History, Bk. VIII*, Pliny the Elder wrote: 'Recently owing to our poverty even bones have begun to be cut into layers ... as an ample supply of tusks is now rarely obtained except from India, all the rest of the world having succumbed to luxury.' He was probably referring to the part of the elephant's tusk which had been embedded in the jaw, as that part was believed to be bone, not ivory. Nonetheless, ivory was becoming so popular that there were some who disapproved of its use, stating that it was an unnecessary luxury, and vulgar as well.

Yet most of the Romans of high rank loved ivory, and, when they could obtain it, they used it for special items. It had probably been used sparingly until the second century BC, when looted treasures from various wars brought back to Rome taught the Romans more about the beautiful material, and influenced their taste.

Among the worked ivory treasures that were imported were chryselephantine statues from Greece. The Romans then copied the technique, though on a much smaller scale. Roman chryselephantine statues were never more than half life-size, and usually depicted emperors rather than gods.

As in Greece, the Romans inlaid or veneered furniture, boxes and chests with ivory or bone. Chairs, couches and thrones were also made using ivory, often with solid ivory legs. Imperial attire dictated that robes were purple with gold thread, and thrones and sceptres were ivory. When the emperors died their funeral couches were made of ivory, among those noted being that of Julius Caesar.

Ivory was also used in the making of curule seats, which had curved, crossed legs and low arms, and which could be folded when the detachable seat was removed. Few people were allowed to use them as they also denoted high status and were therefore only for the use of generals, magistrates, and high-ranking people.

The Romans enjoyed playing board games. Ivory dice, counters, and games boards have been found, many with carved depictions of animals. Less common but also used in Rome were ivory scroll covers. Today all that remains of them are the ivory plaques. Subjects found suitable for carvings included gladiators, warriors, and gods.

Diptychs consisted of two plaques hinged together. They were regarded as very important items, and were given as gifts on special occasions or at important events, such as in celebration of a wedding. They were also made to commemorate

Fig. 3.7 Part of the consular diptych marking the appointment of Rufus Gennadius Probus Orestes. Carved in Rome in AD 530. Height about 35 cm.

triumphs, or to honour the emperor. The most famous use was for the declaration of a newly elected consul. These consular diptychs were sent out throughout the empire. They portrayed the new consul, giving his name and an inscription. Quite a number of these have survived to modern times, although many are damaged or broken. When the Roman politician, Areobindus, came to office in AD 506 diptychs were duly distributed, and six have been found. They are scattered throughout what was then the Roman Empire, acting as proof of the custom.

As the diptychs were sent out for well-documented reasons, it is possible to put a date on many of them. This is extremely valuable as it gives a guideline to styles and customs at the time, against which other, undated ivories can be compared.

Diptychs were carved on the outside, but the inside was recessed to be able to hold melted wax. When it dried, the wax served as a writing block, and the stylus used was normally made of ivory as well.

Until the fifth century, diptychs were carved mostly in Rome. The following century the production moved to Byzantium.

During this time elephants were not only slaughtered for their ivory, but were also used in battle. A company of soldiers rode elephants, just as others rode horses. Those on the elephants were seated in small towers on the animals' backs, and were armed with very long spikes with which to lance the enemy. The elephants themselves also came into combat, locking tusks and using sheer strength to push an opponent away and then attack him with their tusks. It must have been contrary to natural elephant behaviour to act in such a way, and it did not always work as planned. There is a story of a battle where African elephants were confronted by Asian elephants on the opposing side. According to the story, finding the Asian elephants to be larger and fiercer, it is said that the African elephants turned and fled. It is possible that one company of elephants did frighten the other one away, but as African elephants are the larger species it is more likely to have been the Asian ones that fled.

Although ivory was becoming more widespread it was still at this time very much the domain of wealthy people of high rank. It would be several centuries before it became a more commonplace possession.

From the Middle Ages to Modern Times

Late Antiquity and the Early Middle Ages were times of disruption caused by wars. The Roman Empire was split in two, East and West, then the barbarian hordes arrived, and there was constant competition and fighting among the smaller European states. All this brought change, sometimes with different religions, different cultures, and new languages. Ivory use never ceased, but new influences affected styles and usage.

Today there are many examples of mediaeval ivories still in existence in museums and other collections. Some are complete, but many have been broken down or somehow altered. For example, a small plaque may once have been part of a polyptych, and a carving of a Madonna and Child might once have been part of a tabernacle.

In the sixth century Ravenna in Italy was part of the Byzantine Empire. Maximianus, Bishop of Ravenna, commissioned a throne for himself, which can still be seen in the Archiepiscopal Museum in Ravenna. It has a core of wood, completely overlaid with richly carved ivory panels depicting biblical scenes. It is far from being the first recorded ivory throne. For example, pieces from one were unearthed in the ruins of Herculaneum, and King Solomon's ivory throne was mentioned in the Bible, but the throne of Maximianus is the earliest well-preserved ivory throne still in existence.

Many beautiful Byzantine ivory carvings were made in this period. The religion of the region being Christianity, the early ivories were largely for devotional use, but in the eighth century the worship of icons was banned by one section of the Christian Church, effectively stopping their production altogether. Most of those already in existence were destroyed, although a few survived. Fifty years later the rule was relaxed a little, allowing icons to be used once more, although only to be venerated, not worshipped. Early in the ninth century the iconoclasts again had their way, and more images were destroyed. The reasoning behind this was an interpretation of one of the Ten Commandments, stating that graven images should not be worshipped.

Icons were, however, part of Byzantine culture, and eventually the bans were again lifted and icons were permitted once more, allowing the renewed production of carved ivory images of figures from the Christian faith.

The custom of giving consular diptychs was inherited from Rome, and their production continued for a while in Byzantium. Diptychs and triptychs (three plaques held together by hinges) were also made – mostly for devotional use – as well as pyxides (small, round, lidded boxes, usually made of solid ivory without a wooden core), and caskets.

The caskets and pyxides were decorated with religious or secular carvings, so could depict Christian or pagan images, depending on their purpose. Those with pagan images were for personal use, such as for storing jewellery or cosmetics, while a pyxis with scenes from the life of Christ was probably intended for use in church, to hold the sacrament. The caskets were made of plaques of carved ivory, mounted on wood. The plaques were produced in multiples of certain sizes, and mounted on wooden bases made to fit the size of the plaques – an early form of mass production.

Although it continued to be called the Byzantine Empire, the name of its capital city had been changed from 'Byzantium' to 'Constantinople in AD 330, in honour of the Emperor Constantine. Under his rule it had become a powerful trade hub. Constantinople continued to be an important trade hub in the Middle Ages and precious goods flowed into the city. It is not surprising that ivory was usually readily available, although there was a period between the seventh and ninth centuries when it became somewhat scarce, possibly because of Arab conquests in North Africa and the Middle East, which interrupted trade. By the end of the Byzantine Empire ivory had fallen out of fashion, and the centres for ivory carving were to be found further west.

In the eighth century Charlemagne, a Frankish king also known as Charles the Great, came to power. By the time of his death he had been crowned 'Emperor' by the Pope, and he ruled most of western Europe, having accessed large swathes of the continent through wars. His kingdom was called the Carolingian Empire.

Charlemagne spent much time warring with – and winning over – neighbouring states, but he was also extremely interested in the arts and in books – although he could not himself read. There was a Germanic tradition of holding arts and crafts in high esteem, and showing off wealth by presenting valuable and luxurious gifts. Charlemagne adopted the custom, and he also presented gifts to monasteries. Typical items were richly illustrated books with ivory plaques, gems and pearls on the covers, and gem-set reliquaries, some of which were also made of ivory.

The style of the Carolingian ivories is mixed. Most were religious in purpose, so the images represented in the carvings were saints or biblical scenes carved in a Germanic style, but surrounded by decoration taken from Antiquity. Many of the carvings are undoubtedly copied from early manuscripts, and narrative images influenced by earlier religious works were also common, examples of which had been collected for the precise purpose of being copied by the Carolingian craftsmen.

The Frankish Carolingian Empire was succeeded by the Ottonian Empire, which covered the area that encompasses today's Germany, Switzerland, and parts of Italy and France. Carolingian and Ottonian styles are often regarded as much the same, but the Ottonian Empire had renewed the contact with Constantinople, and was open to their artistic influences.

The Ottonians were ambitious and believed that they had a God-given right to rule. Their monasteries produced quantities of excellent and costly works of art, most of which were intended to glorify the Emperor, rather than to glorify

▲ **Fig. 3.8** A pyxis, made in Cordoba, Al Andalus, in the tenth century. The lid and the base are each carved from one solid piece of elephant tusk.

God, although they were destined for churches as well as treasuries. The type of item made was similar to the Carolingian works: book covers, *situla* or holy water buckets, reliquaries, and plaques for furniture or caskets.

Items displayed in museums today are frequently labelled with a suggested country of origin rather than an exact one. This is because the empire crossed so many borders that the styles can be difficult to differentiate, and it can seldom be said with certainty exactly where an item was made.

Early in the eighth century southern Europe – or more precisely most of the Iberian Peninsula and a small section of southern France – was conquered by the Umayyad Caliphate. The area was called Al Andalus and, although now under Muslim rule, other religions were tolerated. The city of Cordoba was declared the capital, and the court grew to be the wealthiest and most cultured in Europe at the time.

For a while there was little or no ivory carving in Al Andalus, but then the custom was brought to the area from Syria. Subsequently Cordoba – and later Cuenca – produced some of the most beautiful ivories of mediaeval times. The carvers depended on wealthy patrons for their work, not on religious institutions. Ivory was carved in the palace workshops, but it is uncertain whether the craftsmen were local or if they came from Syria. Some may even have come from the Fatimid Caliphate in Egypt. It is, however, certain that the work was of an extremely high quality and that ivory – although very valuable – was in plentiful supply.

While the Carolingians, Ottonians and Byzantine carvers were at pains to use elephant tusks in the most economical ways possible (for example, making caskets out of several panels of ivory, and pyxides from the part of the tusk around the pulp cavity with a separate base and top inserted), the carvers in Al Andalus made their caskets and pyxides almost exclusively from a single piece of hollowed-out tusk. This was an extremely extravagant way of carving. As they were made from one piece, the size of the finished boxes was determined

Fig. 3.9 An oliphant, carved in Amalfi or Salerno, Italy, in the eleventh century.

by the size of the tusk and it is clear that some very large tusks were at their disposal, again indicating a good supply of raw material (*see* cutting diagrams, Fig. 2.36).

The pyxides were first fashioned on a lathe in order to make precise shapes where the bases and lids fitted perfectly. The outer shape, hollowed-out centre, and the rabbet (or lip on which the lid fits) were cut by turning. This can be seen on the bases which have a small dot at the centre, left when the block was fixed on the lathe (*see* Section 2.7 Working Ivory: Turning). The pyxides were cut about 1 cm thick to allow for high relief carving. Common to the Spanish pyxides is the somewhat flat surface of the carving, as the carvers started from the outside and worked in, producing patterns in high relief (though virtually never undercutting), but always keeping the outer surface as the baseline from which they worked. It might be called 'flat relief'.

This style of relief was also used on caskets, which it suited well, as the ornamentation of these was often geometric, with possibly a tree as a central point and, on either side of it, a mirror image. The pattern itself depicted animals and birds, and 'arabesques', which were a form of design used a lot in Islamic art and made up of exuberant, scrolling and circling foliage. It frequently covered most of the background. Very occasionally there were hunting scenes which included people, but depictions of human figures were rare in Islamic art. The hunting scenes, and the animals, were often carved in medallions which formed part of the design, surrounded by the foliage.

Unlike further north, where ivory was used mostly for religious purposes, in Al Andalus ivory was used principally to make containers for personal use. They were commissioned by the nobility or by the caliph himself, and often bore inscriptions stating for whom they were made, together with a eulogy to that person, and sometimes a date. Only very rarely did the carver add his name, and then it was always in a discreet place.

Other uses of ivory in Al Andalus were few, but included chessmen and other games pieces, a little inlay, and a few of the usual small, personal items such as bangles or hairpins.

Al Andalus was not alone in making sumptuous items from ivory. Another example of its generous use in the Mediterranean area in mediaeval times is the oliphant. This was officially a hunting horn, made from a whole, hollowed out elephant tusk, but few were actually used for hunting.

An oliphant is an impressive item, very luxurious, somewhat mystical, always fascinating and always held in the highest esteem. There have been many tales and beliefs connected with oliphants, and many old paintings or tapestries depict horn-shaped items that are most probably oliphants. Examples include an early fourteenth century painting by Giotto of the *Adoration of the Magi*, in which one of the three kings is bearing an oliphant, implying that its value was on a par with that of gold and frankincense, and it was a worthy gift for the Christ-child.

Oliphants were often richly decorated in typical Islamic style, although many were made in Norman Sicily. Common to most of them are the decorated bands at both the wide and the narrow ends, with a plain, recessed area in between. A metal band could be inserted here with hooks for carrying the horn, and some had a metal mouthpiece at the narrow end for blowing the horn.

They measured at least 50 cm in length and often more. One style of horn was left plain and undecorated, apart from the carved bands, while another style was slightly facetted between the bands. However, many were carved over their entire surface with the typical arabesques, wild or mythological animals, and foliage of other Muslim carvings. Animals or hunting scenes might be framed in medallions, while in other cases animals chased each other in panels along the length of the horn.

It is thought that the first oliphants were made in Egypt in the tenth century, but the areas from which they more commonly came were Sicily and southern Italy, around Amalfi and Salerno. It is possible that a few were made in northern Europe as well, but there are no dates and very few inscriptions on oliphants, so it is not possible to say with any certainty exactly where or when they were produced. Most were made in the late eleventh to the thirteenth centuries, coinciding with the crusades. As objects to take home from the crusades, they must have been very appealing to the knights, and some oliphants were surely made expressly for this purpose. In a few cases the oliphants commemorated a certain event such as a battle; in others they were a token of the transfer of land.

The secular use of oliphants did not last very long, as they were gifted to churches by the kings, noblemen and bishops of the lands to which they were carried from the south of Europe, notably France, Germany and England. In a few instances the oliphants were put to use by a cathedral or church, and used to hold holy oils or relics of saints. These horns show marks such as holes around the large opening from having had a hinged lid attached. There are also records of oliphants being blown instead of the church bells being rung on

sombre religious days such as Good Friday. In a few cathedrals the horns were put on display, but in many the horns were simply stored in the treasury. It is thanks to this safe-keeping that today there are still about eighty oliphants in existence.

Sicily came under Muslim rule in about 831, when the Christian Byzantine rulers were finally defeated. Islamic rule lasted only until the middle of the eleventh century, when the island came under Norman rule. By the twelfth to fourteenth centuries, though still under Norman rule, Sicily had developed another form of working ivory, apart from the production of oliphants. It was a unique style with decoration which was strongly influenced by Islam. The style has subsequently been called 'Siculo-Arabic', and the main production was lidded boxes, made up of ivory panels mounted on a wooden core. Predominantly round or rectangular, they had lids that were pyramid or truncated pyramid shape. But there the similarity to any other caskets ends, as the Sicilian boxes were held together by metal pins, wires or hinges, and were not carved, but painted. The patterns were the typical Islamic designs of arabesques, foliage, and scenes with animals, all framed in circles or a similar shape, but not usually covering the entire surface as was customary with carved surfaces. The patterns were applied in varying shades of brown pigment, some with a little gilding too.

▲ **Fig. 3.10** Sicilian casket of ivory panels and ebony on a wooden core. The panels have traces of original pigment and gilding. Early thirteenth century.

Although the style of the decoration was Islamic, the patterns did not depict anything religious, so it seems that the Roman Catholics who now bought them had no problem with the designs, though it would appear that in some cases an Arabic inscription was removed.

Styles were different in northern Europe, which had come under Saxon rule following the fall of the Roman Empire in the fifth century. In England, Germanic peoples arrived from continental Europe, heralding the Anglo-Saxon era which lasted until the Norman Conquest in 1066. The Anglo-Saxons excelled in metalwork, but also carved ivory and bone, and in the Middle Ages carving centres were to be found in many of the cathedral cities which were wealthy enough to pay for such luxury.

The ivory used came predominantly from walruses and was imported from the north, and some of the bone used was from whales. The carvings tended to illustrate a mix of religious themes as well as pagan and historical ones, and the decorative parts were the typical twisting and intertwining scrolls of the time, sometimes containing inscriptions written in runes.

Along with ecclesiastical items the Anglo-Saxon carvers produced caskets made of wood with ivory plaques attached, and ivory plaques for book covers. The quality of work was high, and fine relief carvings were produced, in spite of not always having the finest ivory from which to make them. As mentioned earlier, walrus ivory can be challenging to work with, because of its structure (*see* Section 1.4 Walrus *and* Section 2.7 Working Ivory). A few statues were also carved in the round, although this was more unusual and did not become fashionable until later.

In the twelfth century, different ivories were being used in various parts of Europe, all according to what was available and possibly also according to price. For example, in Scandinavia, where walrus ivory was easily obtainable, it was widely used for all manner of articles. Further south in Cologne they were clearly using walrus ivory too, because lots of pieces have been found – along with just a few of elephant ivory – left from an old workshop.

There are many old, ivory chess pieces in existence, but probably none as famous as the Lewis Chessmen. In 1831 a cache of chess pieces came to light on the Isle of Lewis in the Outer Hebrides, Scotland. They turned up in a bag together with some games pieces and a belt buckle, all dating from the twelfth

Fig. 3.11 Three of the Lewis Chessmen, found on the Island of Lewis, Scotland. Thought possibly to be of Norwegian origin. Twelfth century.

Fig. 3.12 The Gunhild Cross, carved in walrus ivory in Denmark in about 1150.

century, and all but a few of the chessmen were carved from walrus tusk. The remainder were made from whale tooth. Their sudden appearance and how they were found remains something of a mystery. Indeed, it is not known when, exactly, they were found, as they may have been discovered earlier but been hidden away for a while. Today there are seventy-eight chessmen accounted for – enough for four incomplete sets.

The tales of how and where they were found are many and varied, and range from a shipwrecked youth who struggled onto the beach carrying just the ivory pieces in a sack; a peasant who found them when the sea washed the sand away from their hiding place, and who thought they were elves and gnomes; a young boy who stole them from a ship and was subsequently murdered; a cow which unearthed the cache with its horns; and some nuns living at a local convent who abandoned them when they left – though there is no reliable record of a convent on Lewis.

However, in the twelfth century there was an important settlement in the area in which they are supposed to have come to light. Although today it is a windswept and somewhat barren landscape, in the twelfth century the Hebrides were part of the Kingdom of the Isles and came under Norwegian rule. It is very possible, therefore, that the chessmen were carved on Lewis by Norwegian settlers, or that they were carved in Norway and were en route further south where they were to be traded. The style in which the chessmen are carved is undoubtedly Scandinavian, for example, the knights are all seated on shaggy ponies that look Icelandic in stature. That they date from the second half of the twelfth century is not in doubt either. But all the rest remains guesswork.

Chess was a popular board game. At about the same time as the Lewis figures were made there were workshops in Dublin that were also producing walrus ivory chessmen. Nevertheless, the European carving centres of the time were usually connected to monasteries, where there was sufficient wealth and knowledge to support them, and where they often enjoyed royal patronage. Also, the monasteries usually had libraries, filled with books written and illuminated by the monks, which could be of inspiration to the ivory carvers.

The most prolific era for ivory carvings in Europe in mediaeval times was the so-called Gothic period from the thirteenth to the fifteenth century. Elephant ivory was expensive (a tusk cost about the same as a peasant's wages for two years), but it was again plentiful, and the carvings were unparalleled in their beauty. Indeed, the Gothic era is considered by many to be the zenith of ivory carving. Among the secular items made, combs were especially popular and were often given as a marriage gift or as part of a dowry.

Gothic carvings almost always told a story, whether ecclesiastical or secular. Furthermore, they often included in some manner the person who had commis-

VENITE BENEDIC
TIPATRISMEI

Fig. 3.13 Diptych depicting (*bottom left*) the Nativity, (*bottom right*) the Adoration of the Magi, (*top left*) the Crucifixion, and (*top right*) the Coronation of the Virgin. Carved in Cologne, Germany. Mid-fourteenth century.

sioned the piece in the illustration. This might be either in the form of the family's coat of arms, or on occasion by depicting a member of the family in a group of people. Thus a lady wearing a wimple sometimes appears in a scene of which she plays no part.

There were many carving centres scattered around Europe at the time. As examples can be named Winchester, Bury St Edmunds, Canterbury and London in England. There was probably a carving centre in Lund in Sweden, but the one in Roskilde in Denmark is well documented. There it was customary to carve in walrus ivory, but more and more carvings were being executed in elephant ivory as it became plentiful. The Gunhild Cross is an example of the local walrus ivory carving of the time (*see* Section 1.4 Walrus).

In Germany there were centres all over the country – the most famous one is possibly Cologne. In Italy, work was produced in the north of the country, as well as other places such as Siena and Venice, and in the Low Countries there was Utrecht – to name but a few. Similarly, in France, ivory carving was carried out in many areas. But the jewel in the crown of Gothic ivory carving was the centre in France's major city: Paris.

All craftsmen in Paris at that time had to belong to a specialized guild. In the middle of the thirteenth century a *Livre des métiers*, or Book of Trades, had been drawn up by a royal official, Etienne Boileau. It contained lists of one hundred and one guilds, plus the rules to which they had to adhere if they were to be allowed to practise their craft.

The guilds were decided largely by what was produced, or the method used, rather than from the raw material used, so the ivory carvers did not have one, single guild but belonged to different ones. For example, there were the knife handles and crucifix carvers; the paternoster, chaplet and rosary carvers (forms of Christian prayer beads); and the *ymagiers tailleurs* or image-makers. Although, theoretically, the carvers kept to their own speciality it seems that there was some overlap. Furthermore, the craftsmen were allowed to work in ivory, and also in other materials such as bone, wood, and horn. The rules were strict, for example, the image-makers were not allowed to work at night, as it was deemed impossible to carry out good work in the dark. Nor were they permitted to make carvings using several pieces of ivory – they had to carve as far as possible from a single piece – as joining pieces together was considered sub-standard.

As with the rest of Europe, the ivory carving production in Paris could be divided into two groups: religious and secular. The items produced for the churches or for private devotion were predominantly diptychs, triptychs, or polyptychs (several tablets, joined in various ways). The most popular motifs were the 'Life of Christ' or the 'Life of the Virgin'. These were depicted in a series of scenes on each tablet, carved in high relief. Tabernacles or host shrines were also favoured, with high relief carvings of biblical scenes surrounding

Fig. 3.14 The Virgin and Child from Sainte-Chapelle. Carved in Paris in about 1260. The Virgin was originally wearing a gilded crown, which has been lost. There are small traces of gilding and polychrome remaining. Height 41 cm.

figures that were carved in the round. Some of the tabernacles incorporated many different materials (for example, wood, enamel, and metals) with ivory as the centrepieces.

Besides the pictorial reliefs, carving in the round gained popularity. The most beautifully carved croziers (pastoral staffs) were produced in Paris at this time. Based on the shape of a shepherd's crook, they frequently depicted the 'Nativity' on one side and the 'Passion' on the other – that is, the birth and the crucifixion of Christ.

Possibly the best-known subject for the time was the Madonna and Child statue, usually with the Virgin standing, holding the Christ-child in her arms (but also with the Virgin seated and the child standing on her knee). Some carvings were large, clearly using most of the tusk, and because of the curve of the tusk there is a sway to the figure of the Virgin so that she appears to be leaning back to balance the child on her hip.

A more unusual item, but a fascinating one, is the *Vierge Ouvrante*, or 'Opening Virgin' (also called a 'Madonna Shrine'). These were mostly carved in wood and painted, but a few were made in ivory. Usually quite large, the statue depicted a seated Madonna and Child, but part, or all of the front of the statue, could be opened to form a triptych, with scenes from the life of Christ on the panels inside.

The secular items produced were of a personal nature: combs with teeth along both sides and carved centre panels; round mirror backs richly carved and with decorative borders; writing tablets likewise richly carved on the back but with a recess on the front for wax; and, of course, caskets. These were made in various ways, sometimes of wood with ivory panels attached, but more usually of solid ivory panels, held together with metal pins, and with hinged lids.

The carvings in the secular items mostly depicted scenes from the tradition of 'courtly love', which was a concept of chivalrously expressing love – usually forbidden love – amongst the nobility, whose marriages often had little to do with emotions and love. The stories were popular at the time and included such tales as 'The Attack on the Castle of Love' and 'The Pursuit of Love'. The caskets illustrating courtly love were covered on five sides with small scenes, sometimes from several different stories, while items like mirror backs were more limited in their scope. Other motifs included courting couples, hunting scenes, jousting, or similar scenes familiar to the owners of the ivory artefacts.

The same type of object was being produced throughout Europe, and often with the same subjects for the decoration. The carving styles varied slightly, but the carvers moved around and took their styles with them, so it is often difficult to tell exactly where a carving originated.

The ivory used at that time was almost exclusively elephant ivory from West Africa. There was no shortage of supply and the tusks were large and therefore adaptable. An exception was a small amount of walrus ivory which was still being made into games pieces.

However, in the late fourteenth century elephant ivory again became scarce. Wars, plague, and disrupted trade routes made it difficult to get a constant supply of ivory from West Africa, so it was used sparingly. Relief carvings were small and carved in low relief, painted, and inset into caskets of metal or other material, like gems, or made into pendants.

Bone carving again became acceptable; indeed, some workshops excelled at the craft of carving bone. For example, the Embriachi workshops in Italy made intricately carved panels in bone, set into the sides or top of wooden caskets, or as panels for wall-hanging mirrors; in both cases surrounded by a mosaic of finely inlaid wood, bone and horn in geometric patterns.

Another style used was incised plaques instead of carved ones. As bone is curved and hollow, thin sheets were cut in order to get the largest possible piece with a flat surface, and the decoration or design was incised and the incisions coloured. The largest bone was whalebone, generally the almost-flat pan bone (jawbone), from which decent-sized pieces could be cut, but whalebone tends to have a slightly grey appearance so was not always suitable.

Small items were still being made from ivory. Paternosters continued to be produced, adorned with a large bead at the end of the row which was often a *memento mori*, reminding the wearer that death is inevitable. These were typically a round bead with a face back and front, one of youth and the other of death in the form of a skull or suchlike.

In the Middle Ages it was the custom to add pigment to ivory carvings. Many, although not all, were 'enhanced' by a little colour here and there – for example, the edge of a cloak or a belt. It was popular to add blue to the cloak of the Madonna, plus possibly some gold decoration around the hem. In relief carvings, the background might be filled in to make the subject stand out more. The skin was never coloured though, as ivory's own lustre was considered a pure and perfect skin tone.

Few of the mediaeval carvings today have their original colouring. The paint wore off with handling, over-zealous cleaning, and was at times deliberately removed when fashions changed. Another reason for its disappearance is the taking of plaster casts, a habit popular in the nineteenth century so that works of art could be seen by more people over a much wider area. When the mould was removed from the carving, most of the remaining colour came away too.

Also in the nineteenth century, it became popular to 'restore' works of art. Unfortunately this was done before real knowledge of the work entailed had been acquired, and anyhow, the Victorians had little regard for whether or not a piece looked original at the end of the process – quite the contrary, they thought they were making improvements. And, because the Victorians liked colour, they added their own, frequently using too much or in the wrong places.

Today the more recently coloured ivories are also faded and much of the pigment has disappeared, making it difficult at times to be able to judge whether the colouring is nineteenth-century or five hundred years older. It is safe to say that it is most unusual to find a mediaeval carving with its original colour. To verify original colour, the pigment has to be tested. This entails taking tiny samples and testing them in a laboratory. The constituent parts will indicate whether they are of the old – and often toxic – type, or later varieties of pigment.

Gold leaf was also a common addition to carved ivories, especially for ecclesiastical purposes. Crowns and halos were frequently made of gilded metal, attached to the carving with small pins. Many have since been lost. If they were carved with the body of the ivory they were usually gilded, as were other areas of statues or reliefs. Again, little original gilding remains, but it may be possible to see where it was from a slight discolouration on the carving, probably left by the glue used to adhere the gold leaf to the statue. Some of the early paints can also leave traces of staining.

Gothic ivories are so special and beautiful that they were also forged, and the forgeries – which were often made in the eighteenth or nineteenth centuries – were also exquisitely carved. Such items are sometimes tested by carbon 14 dating, but as that can only tell when the ivory-bearing animal died (*see* Section 2.4 Testing Ivory – Advanced Testing Methods), it is not very reliable as there were stockpiles of tusks in many monasteries. In some of the best museum collections today there are examples that have the experts in disagreement as to the true age of a 'mediaeval' carving.

Whereas in today's world a fashion can be seen in one country and the image sent within minutes to be copied on the other side of the world, in the late fourteenth century the time taken for stylistic influence to reach from one side of Europe to the other could be a hundred years, as it usually involved the movement of the craftsmen themselves, and the styles therefore travelled little by little across the distances.

The Gothic era was followed by the Renaissance. While northern Europe was still set in its mediaeval ways, Italy was feeling the winds of change and new thoughts were emerging, which were to spread right through Europe. The movement covered everything creative and much that was intellectual too. It affected all forms of art, both fine and decorative, architecture, fashion, music, and science.

Attitudes also changed. More people became interested in science and the arts, and there was a feeling of wishing to improve and no longer simply accepting what was presented to them. However, with the fall of Constantinople and the Ottoman Empire in the middle of the fifteenth century, trade with the East was cut off, ivory supplies were again affected, and ivory lost favour for a hundred years or so.

Deprived of their usual trade routes, the Europeans set forth and discovered new ones. Forty years after the fall of Constantinople, Christopher Columbus had discovered the Americas. Portugal and Spain led the way in discovering new lands and settled in America, Asia and Africa, being followed by the French and the English. Other European countries also followed – although they were not such important players in the field – and meanwhile Russia was conquering its neighbours, and heading for Siberia.

The Reformation brought turmoil in the Roman Catholic Church in many parts of Europe. In some countries the Church came under the governance of the Crown, and the monasteries lost power or were abolished altogether. This meant that, when ivory carving re-emerged, it was no longer produced by the Church for the Church, but was more secular in spirit, and was produced largely by individuals in their own studios, or by craftsmen attached to the courts.

One use of ivory in Europe was for 'miniatures' – small, painted portraits. Being hygroscopic, ivory proved a good base on which to paint in watercolour.

Images of loved ones have always been valued. Before the advent of photography such images were, of necessity, paintings. These were, of course, quite costly and originally only commissioned by royalty or the aristocracy, but by the seventeenth century they had become popular and had reached the middle classes. Originally painted on vellum, it became a fashion to paint miniatures on slivers of ivory instead (*see* Section 2.8 A Few Worked Items).

Another form of miniature portrait was the silhouette. These could be cut in black paper and mounted on white in a matter of minutes, but they were also sometimes painted using black watercolour on ivory.

Slightly further back in time – in the early seventeenth century – starting in Rome and spreading through Europe, the contemporary style of carving was Baroque, which was more fluid and lifelike than previous styles, although in some countries it could also be incredibly opulent and ornate. Figures in paintings or sculptures were anatomically correct, and often only half-clad. A very noticeable difference was that babies began to look like babies, with large heads, and not like the miniaturized adults of the Gothic depictions.

The items made by the carvers were mostly secular, although some things were still produced for the Church (for example, bishops' croziers or altar pieces). Ivory was one of the sought-after materials for the new fashion of 'curiosity cabinets' or 'treasuries'. These were collections of exotic or exquisitely made items, owned by royalty and nobility. In the collection might be silver, gold, amber artefacts, carved rhino horn, and other precious materials.

Typical ivory items produced were plaques carved in low or high relief which could be placed in panels on furniture, or simply hung on a wall. Tall, ornamental cups were also popular for the treasuries. They were made purely for display, and were very elaborately turned, often in several tiers, and often made in a combination of silver gilt and ivory. Tankards, knife handles, and boxes or caskets were also popular items, as were statues carved in the round. The illustrations carved were of gods, *putti*, hunting scenes, and occasionally religious scenes.

Further use of ivory was on weaponry, where crossbow and gun butts were made of wood and inlaid in a white material, although the use of bone or antler was more common for this than ivory.

An example of the increasing secular and religious freedom of thought and action in Europe is the city of Nuremberg, which gained complete independence within the German Empire in 1431. It had the protection of the empire, but did not have to answer to any local princes. It relied on its own council, which ran the city with its own strict rules and regulations. This attracted a lot of wealthy merchants. Also, since it lay virtually at the crossroads of many trade routes, raw materials were plentiful and exporting finished goods was easy, so it also attracted many craftsmen, including ivory carvers. In the mid-sixteenth century there were over five thousand craftsmen recorded as working in Nuremberg, and many became very specialized. The city was also a centre of learning: from 1470 it had printing presses, and it attracted physicians and astronomers.

The city's ivory workers specialized in two things: turned, decorative, lidded cups, and pocket sundials. As in many places, the art of working ivory was passed from generation to generation within the same family, and most of the many sundials that were produced in Nuremberg in the sixteenth century were made by just six family workshops.

Ivory sundials were small, portable, scientific instruments, and were more accurate timekeepers than the early pocket watches. They were also made in other cities (for example, Dieppe in France), but Nuremberg was famous for them, and they were recorded in the city's records from the end of the fifteenth century. The sundial makers were called 'compass makers', because the instruments relied on a compass to calculate the time (*see also* 2.8 A Few Worked Items).

The lidded cups for which Nuremberg was also famous were spectacular items that could stand 50 cm tall, or more (*see* Section 2.7 Working Ivory: Turning). The craftsmen at Nuremberg were masters of turning, making objects

◄ **Fig. 3.15** The Danish coronation throne, made in the mid-seventeenth century, using narwhal ivory.

ever more spectacular in their execution. They produced extremely thinly cut ivory – so thin that a light could be shone through it – highly ornamented and in various profiles, not simply round ones. (Another centre that excelled at turning ivory was Dresden.)

At this time ivory was being used for furniture production. An extravagant example of the use of narwhal ivory was the Danish coronation throne, made for King Frederik III in the mid-seventeenth century. It is made up of many whole tusks, plus several shorter lengths of tusk and flat sections of many more (*see also* Section 1.6 Narwhal).

Ivory complemented other materials. It could, for example, be used in combination with amber. On caskets, strips of ivory formed less brittle edging and corners, as well as keyholes, and, as decoration, carved plaques were inset in the amber, and finials added to the top. On furniture it could be mixed with other materials such as various woods, mother-of-pearl or tortoiseshell as marquetry. A master at the art of decorative marquetry on furniture – especially illustrating flowers and plants – was the cabinet maker, Pierre Gole. He was born in Holland in 1620 but moved to Paris at a young age. He became Master Furniture Maker to the King in 1656 and carried out a lot of work for Louis XIV at Versailles and other royal palaces.

▲ **Fig. 3.16** Cabinet with marquetry of ivory, dyed bone, tortoiseshell and various woods. Made by Pierre Gole in Paris, in 1665.

Following the Russian conquests of Siberia in the sixteenth and seventeenth centuries, Russia took over the export of raw walrus ivory, shipping it from the new port of Archangel on the northern coast. Records show that they also exported 'fish's teeth', presumably meaning whale teeth.

The Russians had been importing walrus ivory from Siberia since the eleventh century, or possibly even earlier, and using it for a form of cottage industry carving for domestic consumption. Mammoth ivory was also available and carved alongside walrus ivory, although it was less favoured as it was more expensive. In 1694 ivory became a state monopoly under Tsar Aleksei Mikhailovich and the work was refined. It lost its rustic style and became more European, and only within the reach of courtly people.

There were two main centres for carving: Kholmogory, on the Dvina River not far from Archangel, and Moscow. The workshops in both areas carved both bone and ivory. The carvers in the northern city tended to carve subjects more related to their historic work, such as animals and human figures, and often these were stained green or brown. Meanwhile in Moscow the concentration was more on chess sets, ivory plaques for sewing boxes, and other luxury items. Most of the finished work continued to be for domestic use, and unworked tusks constituted most of the exported material.

Carving in Russia reached its peak in the eighteenth and nineteenth centuries. At this time more finished items carved in walrus and mammoth ivory (and also in bone) were being exported than previously. The designs included openwork relief, surprisingly finely carved in walrus ivory, which can be tricky to work with. Popular items for export included boxes, games pieces, and handles for teapots or knives. Jewellery was also made, some comprising necklaces of multiple finely linked ivory chains. Unworked, raw ivory tusks continued to be exported as well.

The ivory of choice in most of Europe was elephant ivory, which was again arriving from both Asia and Africa. Walrus ivory, with its central core of secondary dentine, would have been too limiting for the type of work that was being executed. In Scandinavia, especially at the Danish court, narwhal ivory was utilized, although its long, thin, spiralling shape and the fact that it is hollow were restrictive. However, sections encasing a silver core made decorative tankards and, as already mentioned, the Danish coronation throne was made of narwhal ivory (*see* Section 1.6 Narwhal). As there was still much debate at the time as to whether narwhal ivory was, in fact, unicorn horn, the throne attracted much attention and admiration throughout Europe.

In other parts of Europe narwhal tusks were used for the staffs of bishops' croziers – an item for which they were ideally suited, being long, narrow and straight, and of an attractive and very valuable material. The crooks at the top were carved from elephant ivory.

By the eighteenth and nineteenth centuries the use and possession of ivory was no longer confined to the courts or aristocracy, or to the Church. The Industrial Revolution gave rise to the middle classes. Anyone with sufficient money could buy ivory, and a whole new world opened up for the carvers. In some places ivory carving almost ceased, while new centres appeared elsewhere. The carvers opened their own shops in places where they could sell their wares to wealthy merchants, and the genre of items produced changed.

Little work was being carried out in Germany, and the production was still firmly rooted in the Baroque style, with the few centres still under the patronage of a few wealthy benefactors. Some carvers began to change to a more Rococo

Fig. 3.17 An Erbach rose. Height 2 cm.

style, and in centres such as Dresden some beautiful and gentle figures in natural poses were produced, along with many small items for personal use, such as toilet articles.

In the middle of the eighteenth century Count Franz I of Erbach-Erbach, in the Oldenwald in Germany, was sent by his family to travel around Europe on his Grand Tour, as was the custom amongst the aristocracy of Europe. Whilst in Italy he saw ivory being carved, and decided that his people back home – who were mostly wood carvers – could easily learn the craft. He sent a few of them to Italy to be trained, and in 1783 he founded an ivory workers' guild. The work produced was shown at exhibitions, Erbach became well known, and orders poured in from around Europe.

Count Franz was one of the courtly people who owned a lathe and worked ivory himself as a hobby. This was a popular pastime among the aristocracy, many of whom enjoyed the art. Although simple, the items produced by Count Franz – such as tobacco boxes – showed a high degree of skill.

Erbach lies in a forested area, where there had always been hunting. Originally the main carving themes were hunting scenes, deer, wild boar, hunting dogs and so forth, carved in the round, the figures sometimes mounted on boxes. Jewellery had similar themes. Over time these subjects were added to and many and diverse figures were produced. Erbach became known for its jewellery depicting roses, and indeed they were sold as 'The Erbach Rose'.

Today there is an ivory carving school in the adjacent town of Michelstadt, where students can take a three-year course, starting work on bone and progressing to mammoth ivory as they become more accomplished. They also carve Elforyn, a synthetic material made predominantly to imitate ivory.

In France ivory had largely lost its popularity. Louis XV preferred a more romantic style of decoration and ivory was no longer as fashionable. Other media took over, especially porcelain, which was very versatile as it could look as simple and pure as ivory, or colourful, ornate and fussy, depending on what was required.

The French ivory carvers were still split into their various guilds, so they did not form a strong body of craftsmen, and the studios tended to move away from the main cities. A few carvers remained in Paris, producing various small objects, the most common of which were crucifixes for private devotion.

One centre, however, was thriving. Dieppe was a coastal town, which had long been the port where the ivory arrived from Africa, and from where it was sent on to the workshops in Paris. With so much ivory so readily available it became an obvious place for a carving centre to develop. Some records indicate that there was some carving being done in Dieppe in the fifteenth century, but the general belief is that it started towards the end of the seventeenth century.

When other centres closed, Dieppe carried on. After the town was partly destroyed by the Anglo-Dutch bombardment in 1694, Dieppe was rebuilt and the ivory carvers resumed their work. The carvers were individuals without royal patronage, who set up their workshops and sold to the local merchants. It was a wealthy town, and the demand for ivory was good. Further, it was greatly improved in the 1820s when Dieppe became a fashionable seaside resort, attracting people from Paris and from England too. Good-quality 'trinkets' of fine workmanship made extremely popular souvenirs.

Sometimes, when Parisian carvers had too many orders from their customers, the carvers in Dieppe supplied some finished items or missing parts. The figure of Christ on the Cross was a very popular item to buy at the time, and was an example of this teamwork. The figure was of ivory and the cross of dark wood. Because of the shape of the figure it was always made from three pieces of ivory, the arms being separate from the body. So many crucifixes were being ordered that Paris often could not keep up with demand. The figures were made in standard measurements, and for a while Dieppe was supplying most of the arms for the Christ figures which were 'made in Paris'.

Dieppe had its share of studios and workshops that were kept in the family and passed down the generations, but in 1808 a school was started in the town, where the art of ivory carving was taught to, among others, youngsters from a local orphanage.

The carvers in Dieppe were expert at incredibly fine carving. They specialized in ships, with paper-thin sails and thread-fine rigging, all made from ivory. Another of their specialties was their *mosaïque* relief, a form of pierced relief as fine as lace, which was incorporated as background to the main subject of the carving. It was commonly used on the lids of small boxes or as the shield for a candle. Other items for which Dieppe became known included tobacco graters, statuettes, *bénitiers* or holy water holders (sometimes with a real seashell as the bowl), sundials, and framed scenes carved in pierced ivory and mounted on a dark background to give a silhouette effect.

Also in France, in the early nineteenth century a piano-maker by the name of Jean-Henri Pape invented the method of sawing very large, fine sheets of ivory, which in theory made it possible to veneer furniture with large sheets instead of piecing together a lot of smaller ones (*see* Section 2.7 Working Ivory.) However, the ivory sheets were so thin that they were somewhat transparent, so the method was only suitable for using on furniture made from very pale wood. The method did, however, benefit the ivory miniature painters, who were able to be more economical with their ivory, and to make much larger pictures if they so wished (*see* Section 2.7 Working Ivory: Miniatures).

In the 1850s a patron of the arts, the Duke of Luynes, commissioned a sculptor by the name of Pierre-Charles Simart to reconstruct Phidias's chryselephantine statue of Athena, which had stood in the Parthenon in Ancient Greece, for his chateau. Silver and bronze were to be used, rather than gold, and it was to be a lot smaller than the original and stand less than three metres tall. The commission was executed and the statue exhibited in Paris in 1855 to some acclaim, although there were some critics who remarked that it had cost the duke 100,000 francs to prove that Simart was not Phidias. Simart worked mostly in marble, so the chryselephantine statue was a new venture. However, the statue gained much attention and helped to spawn a renewed interest in chryselephantine work. Although this was not immediate, fifty years later several artists were making statuettes in ivory and bronze or other metals, which fitted well with the Art Nouveau movement.

At that time some extremely beautiful jewellery incorporating ivory was also appearing. Mixed with precious metals and gems or enamel, the Art Nouveau jewellery was in a class of its own. With its soft shapes and flowing lines it was both lifelike and mystical at the same time. The master of the craft was a Frenchman, René Jules Lalique.

Little artistic carving of note was being done in England at this time. A few items were made here and there, but the mediaeval centres of excellence were long gone. There was, however, one man who worked on small ivory busts and statues, copying them mechanically from larger models. Benjamin Cheverton was able to make almost exact miniature replicas of, for example, stone busts. Although it was somewhat more complicated, his 1836 machine was akin to today's machines that copy door keys.

Ivory was also being used in the world of medicine. The wealthier doctors and dentists often used instruments with ivory handles, which – although they looked good – cannot have been very hygienic. Ivory dentures were popular with anyone who could afford them, the alternatives being a selection of other materials ranging from vulcanite (rubber cured with sulphur) to second-hand human teeth. Hippo ivory was the best for dentures as it was hard and did not stain easily. Elephant ivory was deemed a little too soft for the purpose and very expensive, but for less expensive dentures walrus ivory was used, with the front teeth carved from the primary dentine, and the palate and back teeth from the globular-looking secondary dentine.

In 1891 a German doctor by the name of Glück used ivory prostheses for hip joint replacements, and experiments followed using ivory prostheses for the wrists and other joints as well.

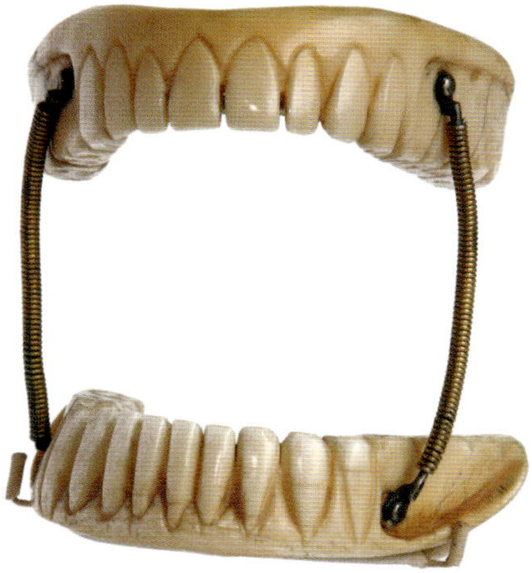

Fig. 3.18 Hippo ivory dentures, held together with piano wire.

Another use of ivory in relation to teeth was as teething rings for babies, as the smooth surface yet hard material was a satisfying medium for a baby to chew on to lessen discomfort.

At this time ivory objects were being imported into the UK from India, China and Japan, and artefacts from Africa were arriving in museums and private collections. They were different, and they attracted much admiration and attention. Meanwhile in Europe – with a few exceptions as noted – the production of ivory goods was more about utilitarian objects than decorative ones, and it was being used for furniture inlay, brushware, toilet articles, small boxes or pots, cutlery handles, page turners, pepper mills, and other useful objects.

Fig. 3.19 Edwardian baby's silver rattle with ivory teether. Made in England, 1906.

During the eighteenth and nineteenth centuries vast amounts of elephant ivory were being consumed in the production of just two items: piano keys and billiard balls. London was the second largest producer of piano keys after New England, USA. They were made in specialist piano factories (*see* Section 2.8 A Few Worked Items).

Other factories produced the billiard balls for what they termed 'cue sports' – a name which covered various games played with a wooden cue and balls on a cloth-covered table. The name 'billiards' tends to be used generically to cover all the different games.

Billiards dates back to the fifteenth century, and the balls were then made of wood. The first mention of ivory billiard balls is in the inventory of an English nobleman's house in 1588. Between the seventeenth and twentieth centuries, ivory became the material of choice for billiard balls. They had to be cut and turned on a lathe, and allowed to mature before being turned again. If

the process was not completed correctly, the balls would be uneven and would not roll correctly. Storage of finished ivory billiard balls was also crucial because they could alter in too dry or too damp an atmosphere. They seldom lasted very long in perfect condition, so had to be replaced. The use of materials other than ivory was, as with the piano keys, met with opposition, as the other materials neither reacted nor sounded like ivory (*see* Section 2.8 A Few Worked Items).

▲ **Fig. 3.20** Backgammon pieces with pink- and green-dyed shagreen surfaces, and elephant ivory surrounds. Made in England, about 1935.

All production of ivory items ceased in Europe with the complete bans on the trade in ivory brought in by CITES (*see* Section 1.1 Conservation). In some European countries, such as Great Britain, it was interpreted as a total ban on the trade, including the domestic markets. In other countries, pre-existing stockpiles can still be used and items sold within the country, but only under strict controls of the appropriate authorities. Thus the ivory school in Michelstadt occasionally allows its best students to work in elephant ivory from its pre-ban stockpile, and in other places such as Dieppe it is still possible to find one or two ivory carvers at work who had a good collection of the raw material when the bans came in.

New ivory is still being carved and sold in some other European countries. This may be from a permitted quota, but would need a permit for export. An example is Russia, where mammoth and walrus ivory (and bone) are still carved and can be found in shops in Moscow and other large cities, where they are mostly sold to tourists, although walrus ivory needs permits for importation to most other countries.

▼ **Fig. 3.21** Pocket sewing set, made from two solid pieces of ivory. Made in England, 1908.

Museums are also allowed to use ivory to repair or renovate items in their collections. Many museums have small stockpiles of raw ivory, although sometimes they need help to get the right material. A few years ago it was decided to renovate the Danish coronation throne made of narwhal ivory, but the museum where it is housed, Rosenborg Castle, did not have the necessary tusks. An appeal went out to the general public requesting the donation of tusks from private collections, and enough was offered to complete the renovation.

3.5

Egypt and the Middle East

Egypt

It is known that ivory was used in Pre-Dynastic Egypt, 5500–3100 BC, and continued to be used throughout all the dynasties. However, ivory does not keep well when exposed to the dry heat of Egypt, so what little remains has been found in burial chambers. Sadly, most of these have been plundered, so very few actual examples remain to indicate exactly how it was used, although from the paintings on the walls of burial chambers it is possible to deduce a fair amount.

Among the artefacts found are examples of bangles, hair combs and pins, spoons, ladles, jars which probably held cosmetics, and harpoon tips, which date back to the Badarian Period, about 4500 BC. A thousand years later the Egyptians were carving ivory handles which were mounted on flint knives, and while the flint blades were roughly chipped, the ivory carvings were quite intricate. Games pieces carved with pictures of lions, dogs and monkeys have been uncovered from the Temple of Abydos, and dated to about 3000 BC.

Ivory was only available to royalty, and it was regarded even by them as valuable and luxurious. It was an obvious choice for taking to the afterlife, and ivory items in tombs included small statues, headrests, and the magic, ivory wands that enabled the deceased to use the various items that had been put into the tomb to accompany him on his final journey.

Some of the ivory was local. At that time there were elephants in Syria and the surrounding countries, and hippopotamuses in Egypt. There is a record of Thutmose III hunting elephants in Syria, and the Syrian elephants are believed to have died out some time during the first millennium BC. It is unknown whether the species died out purely because they were killed for sport and for their ivory, or whether other factors were also involved.

The Ancient Egyptians were traders and imported goods from other countries. These included cedar wood from the Lebanon, lapis lazuli from Afghanistan, and gold from Nubia. From further south on the African continent they imported ebony and some elephant ivory, although as long as there were hippos to be found locally, they carved mostly hippo ivory. The use of elephant ivory appears to have increased as the numbers of hippos decreased, and elephant ivory was possibly the most common type used by the Eighteenth Dynasty (about 1550 BC). During this period Egypt enjoyed great wealth. There are records of Queen Hatshepsut sending ships down the Red Sea to Punt to bring back ivory and incense, and luxurious ivory items were carved at Thebes for the use of the nobility and royalty, and also to be used as gifts exchanged between the kings of different areas. It was the custom to request gifts, not to be offered them, and the gifts were lavish. Tributes were sent to neighbouring courts and threats of war were occasionally issued for non-payment.

The ivory workers of Egypt were masters of the art of carving, and were also able to use every method of treating ivory that is known today, albeit using far more primitive tools and often very toxic substances. The pigments used for colouring were especially toxic. Carving was traditionally men's work, but colouring and staining ivory appears to have been done by women.

The tomb of Tutankhamen is famous because it was still intact when it was discovered in 1922, about three thousand years after the king had been buried there. It had been left totally undisturbed, and was found to contain many very precious artefacts. Among them were a few items of ivory such as a headrest, made from two pieces of ivory and joined with a wooden dowel through the figure of the god Shu in the centre. It had been etched and coloured with blue pigment, probably made from ground lapis lazuli.

Other ivory objects in his tomb included a pair of 'clappers' and a boomerang. Clappers were a musical instrument comprising two slices of ivory carved as hands at one end and joined together at the other end, so that when knocked the hands 'clapped'. Both the boomerang and the clappers are sharply curved, suggesting that they are made from hippo incisors. The simple ivory bangles made in two halves and hinged together are probably also hippo ivory.

Fig. 3.22 Ivory headrest said to come from the tomb of Gua. The openwork shaft is in the form of two *tjet*-symbols or 'Isis Knots'. Egypt, Middle Kingdom Period.

Lacking cupboards or other furniture for storage, boxes and chests were prevalent in Egypt, as in many parts of the world. Several wooden boxes were found in Tutankhamen's tomb, some covered in various materials including strips of ivory incised with hieroglyphs which were stained with pigment.

In Roman times Egypt was ruled by the Romans from Alexandria, and by the end of the sixth century Egypt was part of the Byzantine Empire, having been fought over by various armies in the interim. By the late tenth century, the Shia Ismaili caliphate of the Fatimids had made Egypt their centre, with Cairo their capital.

All these different cultures brought with them their own decorative styles and ways of using ivory. The styles tended to merge as they were developed to suit the region. For example, traditional Ancient Egyptian motifs might be combined with depictions of Greek gods, while the later Muslim influence introduced vegetal designs and arabesques. Each motif was accepted on its artistic merit rather than for any political or religious reasons.

There is little ivory remaining from Fatimid Egypt, but what remains is of the highest quality. Hunting scenes and other courtly pursuits were carved in very high relief on ivory panels, with the animated figures in the foreground and the foliage receding into the background.

It is thought that some oliphants – ivory hunting horns – were made in Egypt in the tenth century, and that the custom started there (*see* Section 3.4 Europe). It was an extremely luxurious form of ivory carving, as it used a whole elephant tusk which was hollowed out. Oliphants seldom had any inscriptions, and in the rare cases that they did, there was never any indication of where the horn was made. Research is still ongoing.

By the time the Mamluks reigned from Cairo in the fourteenth and fifteenth century ivory was being used almost solely for inlay in mosque furniture. The inlay was in the form of small panels, either carved or left plain, arranged in geometric patterns.

Egypt had long been an important hub on the trade routes, and for centuries raw ivory had passed through Alexandria before being shipped to various ports in the Mediterranean. In Roman times it went to all the outposts of the Roman Empire, and although trade was disrupted for a while during the mediaeval period because of changes in alliances, it was encouraged by the Mamluks who were excellent traders, and elephant ivory again flowed into Europe.

The ivory carvers in Egypt were originally from specific families or tribes and were often Copts. With the change to Islamic culture the Copts continued their work, living in small groups in the largely Muslim communities. It is thought that some formed ivory workers' guilds.

The Middle East

The custom of carving ivory appears to have been taken east from Pre-Dynastic Egypt where it started. In all areas the material was used predominantly to decorate furniture or boxes, with some being made into small personal items such as jewellery, bangles, pins, jars for unguents or cosmetics, or mirror and fly whisk handles.

Ivory was for royalty and nobility only. It was luxurious, expensive and sought after, and was frequently amongst the precious items looted as war booty. In some of the palaces that have now come to light there are hoards of ivory which appear to have been stored. Such abundance of ivory in one place points more to the spoils of war than to extravagant commissions for the palace in question.

As in many areas of the world, the borders and names of the Middle Eastern countries have changed a lot in the past 600 years. Mesopotamia – the area lying between the Tigris and Euphrates Rivers in modern-day Iraq – covered Sumer, Akkadia, Babylon and Assyria, and is considered to have been the cradle of Western civilization. Little was known in the past about these kingdoms and how the people lived their lives. With the fall of the Ottoman Empire at the beginning of the twentieth century, the doors were opened to archaeologists and many excavations were started. The First World War interrupted the work, but it resumed in the 1920s and 1930s, only to be interrupted yet again by the Second World War, and then also by other, subsequent wars in the regions. Fortunately, many of the digs have been revisited and work is still being carried out whenever possible to learn more about the history of that part of the world.

At the beginning of the twentieth century finding such sites was rather hit-and-miss, and relied on anecdotes and anomalies in the landscape. A lot has been learned from the written tablets left behind, in scripts such as cuneiform. Today it is possible to use photographs taken from the air, or indeed from space, to pinpoint areas that look as though they may be hiding ruins.

In the centuries before Christ, the majority of the sculptures and carvings from the area were cut in stone, but ivory was also used as it was held in high esteem, so throughout the region there were centres where ivory could be carved. Most carving – both in stone and in ivory – was executed in relief, and very little was carved in the round.

Very broadly speaking, when decorating furniture, the Sumerians favoured applying cut-out pieces of ivory, while the Assyrians cut plaques and incised patterns and pictures in the plaques. The Phoenicians carved mostly in low relief. In all areas a lot of pigment was used and the finished results were probably quite colourful. Much of the carving was also overlaid with gold.

Many graves have been found in Ur, an ancient Sumerian city state in southern Mesopotamia which was first founded in about 3800 BC, although the archaeological finds are from much later. Amongst the items found were various games boards and pieces, some inlaid with ivory. Mirror backs, small pots for cosmetics, and bangles were also typical items for which ivory was used.

An exceptional find was made near Beersheba in what was then Palestine in the 1950s, of a troglodyte village. One of the underground rooms was an ivory workshop, complete with tools, workbench and raw tusks. The ivory has been carbon dated to about 3300 BC.

Other finds include about three hundred pieces of carved ivory at Megiddo, dating from the Late Bronze Age. There were many different types of item, and as Megiddo lay on the trade route between Assyria and Egypt it is thought that the items were not made locally but were imported, and that they represented the king's treasury which had been hoarded as a form of wealth.

A few ivory carvings have been found in the ruins of Nineveh, and in Samaria, which is now in Israel's West Bank. Many other places have also revealed small finds, but the most important and by far the most famous were ivories found at Nimrud in Assyria, which lies about 30 km south-east of Mosul in today's Iraq. The palaces at Nimrud were built in about 900 BC for Ashurnasirpal and his son, Shalmaneser III. They stood for about two hundred years.

Digging started in the middle of the nineteenth century, and for a long while the site was thought to be much like all the others – extremely interesting but with nothing to single it out. No ivory at all was found to begin with, but slowly small caches emerged, found in various areas of the archaeological dig. At that time the site was thought to be part of the Nineveh site. In 1855 a room with stored furniture covered in ivory had been found and was considered a fantastic result, but more was to come.

The Nimrud site was revisited several times by archaeologists before they finally finished in 1963. It was towards the end of the investigations that the largest cache of many thousands of ancient ivory carvings and pieces thereof ever found was discovered. They were in a dried-out well, and it is thought that

Fig. 3.23 Ivory panel from the Palace of Ishpurnal at Nimrud, showing a male human-headed winged sphinx walking amongst flowering plants. He wears the double crown of Upper and Lower Egypt. Carved in about 900 BC.

they may have been hidden there for safe keeping when an invading army approached (*see also* Section 2.6 Care and Damage). Another possibility is that they were stripped of their gold overlays and discarded by the invaders, who either decided that they could not carry all the ivories back home with them, or possibly did not want to, because ivory was not fashionable with them at that time.

The number of ivories discovered in the ruins of Nimrud is amazing, especially considering their age and the fact that the palace was sacked. There is a clear Egyptian influence in some of the carvings, indicating strong ties between the kingdoms. Others show pure Assyrian style, and yet others are more Phoenician. The differing styles suggest that they were not all carved at Nimrud, but that at least some of them were spoils of war.

The favourite motif was people, in scenes of hunting, war, or enjoying some sort of ritual. Also popular were bulls, birds, and winged sphinxes. Some of the ivories show signs of having once been painted, while others have evidently been in a fire and are completely blackened.

After the fall of the Assyrian Empire in 612 BC, Mesopotamia was ruled by a succession of foreign dynasties. In 332 BC it fell to Alexander the Great and became part of the Greek Seleucid Empire. Later the Parthians and the Romans fought for control of the area, and at the beginning of the third century AD Mesopotamia came under the Persian rule of the mighty Sassanid Empire. By the seventh century it succumbed to the spread of Islam and ultimately became part of the Ottoman Empire.

However, trade with surrounding countries continued, and ivory continued to be purchased as a luxury item. It was brought in from various places, including the Indian subcontinent with which there were strong ties. Not all the ivory brought in was in its raw state to be carved locally; indeed, it is thought that some was probably worked in nearby Phoenicia. There is also the probability that some ivory – both raw and worked – was obtained by looting.

As always, the conquering empires brought with them influences in customs, styles and use of ivory, as also did the countries with which they traded. Earlier art in the area was influenced by Classical and Late Antique Mediterranean styles, while in mediaeval times the influences came from Europe. An example of exchange of ideas is board games, which were popular at court and amongst the nobility. The Indian subcontinent brought chess to the Persians (today's Iran), while the Persians brought backgammon back to them. Chessmen were traditionally carved in ivory – one of the few, early examples of ivory carved in the round.

With the advance of Islam came the Islamic styles. In sharp contrast to Christianity, Islam does not permit religious images, so while ivory was mostly used for ecclesiastical purposes in the Christian lands and commissioned by the Church, in Muslim countries the use of ivory was mostly secular, and paid for by the court, or by people of great wealth and high rank.

The Islamic style of roundels or medallions is seen on many things including textiles and paintings. Repeating patterns of scrolling foliage made up of vines, palmate leaves and sometimes grapes and flowers, encircled figures or animals and birds. They are called 'arabesques', and are prevalent in carved ivory decoration. Although no religious form could be depicted, figures in courtly pursuits were acceptable, as were the animals which were regarded as 'creatures of God'. Where such scenes were usually depicted in architectural surroundings in European carvings, the Islamic figures always had vegetal surroundings. These basics continued to be used for several centuries in the Muslim world.

In the seventh to eighth centuries small, round boxes called pyxides were being produced in Syria. A few may also have been produced at this time in Egypt. They had a typical, bulbous style, with domed lids, and were carved in low relief in what appeared to be two planes, which has been referred to as 'flat relief'.

In the Middle East there was a lull in ivory carving when, in the thirteenth century, the Mongols overran large swathes of the region. The Mongols' capital was Samarkand and although renowned for the destruction of everything in their paths, they also produced beautiful carvings in many materials. However, there is little evidence that they used ivory.

In about the fourteenth century, the Persians developed a very intricate form of inlay called *khatamkari*, which gave the effect of tiny pieces of contrasting woods and ivory inlaid onto the surface in rich, geometric patterns. It was used to cover doors and furniture (*see* Section 2.7 Working Ivory).

The Ottoman Empire was expanding, and in the mid-fifteenth century the Byzantine Empire came under their rule and Constantinople became the capital of the whole Ottoman Empire. From then on ivory was used almost exclusively as inlay, with very little used for personal items.

Exceptions were, however, to be found, for example, sword and dagger hilts, as well as a few other items such as belt buckles, and pen holders which were carved in relief. Walrus ivory was used alongside elephant ivory, possibly because it was readily available through trade, but possibly also because – at least for the hilts – it was considered slightly more porous and provided a better grip.

The Ottomans liked colour and all surfaces were richly decorated. Inlay was used as adornment in mosques, where the *minibar* (pulpit), as well as prayer niches, low tables, and doors might all be covered with wooden panels filled with geometric inlay.

Typically a panel consisted of a star with an even number of points as the central design. From the star radiated triangles, polygons and wedges in strict geometric patterns. The sections were constructed from pieces of wood and the recesses inside each section were inlaid with contrasting wood, ivory or bone. The small ivory plaques were sometimes left plain, but could also be carved in relief with the usual Islamic designs of vegetal decoration and arabesques. Sometimes the whole design was taken further and the geometric patterns were themselves surrounded by arabesques, often using mother-of-pearl in the design to enrich the effect.

From the sixteenth to the eighteenth centuries the Safavids in Persia had strong links with the Indian subcontinent. The Safavids were patrons of the arts, and collectors too. They produced some fine ivory work themselves, but it is thought that a lot was imported from India.

In the nineteenth century the whole area became less insular and started opening up to the rest of the world. At the same time, in Europe, attention was being drawn to exotica, and items from strange and foreign lands were in demand. Consequently, furniture was produced in European style, though with ivory inlay. The greater demand meant that less attention was given to individual pieces and the quality began to decline.

The production of *khatamkari* was restarted, but used on small objects that might tempt the new influx of tourists, and seldom on doors and furniture as before. Simple ivory marquetry was also made, and in the next century production concentrated more on inlaid cigar boxes, trays and board games, produced for quick sale to visiting tourists. At first the work was of a reasonable standard, but as more tourists visited and more and more items were produced and sold in the bazaars of the Middle East, the quality fell. Ivory and mother-of-pearl ceased to be used and stained woods or bone were used instead. The production continues today but, with some exceptions, the patterns are no longer as complicated and the style is a poor imitation of the earlier, beautiful works.

3.6

Africa

In many African cultures elephants are regarded with some reverence, and no other animal is afforded such respect, and no others are the subject of so many stories and myths.

In some areas elephants have been thought to be capable of metamorphosis, changing themselves into other shapes or beings. In a few stories, mankind is descended from elephants. To carry something that has belonged to an elephant – for example, an item made of ivory – imbues the wearer with some of the power of the animal.

Elephants were regarded as peaceful, social and caring, which were the best human traits. They could also be wild, angry and dangerous, which were also human traits. Further, they were long-lived, which indicated wisdom. All these characteristics were believed to be carried in the ivory that came from these large beasts.

It is not surprising that elephants feature widely in African art, even in the areas where the people have never set eyes on one. Their tusks, both carved or left in their natural state, were held in great esteem and used mostly for ceremonial purposes. As in other regions, ivory was a symbol of status – although the reasons in Africa went much deeper.

African peoples very seldom killed elephants solely for their tusks. If they did, it was in order to give the tusks to their chief as tribute. Elephants might be hunted as food, as the meat from an elephant could feed a whole village for a long time. In some African cultures – for example, among the Masai – it was a rite of passage for a young man to kill an elephant. The animals were big and powerful, and before the use of firepower they were extremely dangerous to hunt, so it was necessary for a young man to show how brave and strong he was. In other African cultures the animals were never hunted. Among the people of Togo it was believed that dead elephants brought bad luck, and that they should not be touched, even to remove the tusks.

In most African societies one of the tusks from a dead elephant automatically became the property of the king or ruler. Usually he received the tusk that was nearer to the ground when the animal fell, as it was thought that it must be the larger and heavier of the two.

The amount of worked ivory that can be dated with certainty to before the arrival of the Europeans in Africa is relatively small. A few carvings have been found that are known to date back to Neolithic times, and some elaborately carved ceremonial regalia from the tenth century was found at Igbo Ukwa in Nigeria. In Ghana some hippo ivory artefacts thought to be from between 1300 and 1650 have been found where there was once a village, which gives an indication of how uncertain the ages are. Most items found – even later ones – are of indeterminate age. It is also difficult to know exactly how items were used. Every item had a symbolism, but the exact meaning has in many cases been lost.

Each society had its own way of using ivory. Not many made utilitarian items, as the material was considered too special. In some parts of the continent pendants were put on their cattle, presumably as amulets and probably to protect them from predators or to increase their milk yield. Ivory throwing sticks were used in Kenya amongst the Masai. As adornment, ear-plugs, finger rings, and symbolic carved beads strung in necklaces together with coloured glass beads were also seen in various parts of the continent, and small ivory discs were used as currency. Whole unworked tusks often surrounded the throne of a chief or king, and the ancestors were honoured by placing whole tusks on alters or shrines.

Ancestors were of great importance in all African societies. They bore wisdom and were to be honoured, and they were consulted in times of difficulty or danger. As with the wearing of elephant ivory as a means of transferring the power of the animal to the wearer, it was also thought that the wisdom of deceased ancestors could be carried in their teeth, so in some regions it was the custom to extract the teeth of an elderly relative when he or she died, and wear the teeth strung as a bracelet.

Much of our knowledge of African ivories comes from the fifteenth-century European traders who discovered the locally carved items and sent them back home, where they were stored first in royal treasuries as curiosities and later in museums.

There are many Afro-Portuguese ivory items in the museums in Europe and North America which date from the sixteenth century, when the Portuguese arrived. They, and other Europeans who followed them, set up trading posts in Africa (*see* Section 3.1 Raw Ivory Trade Routes). As in other parts of the world in which they settled, the Europeans commissioned work by the local craftsmen, and influenced their styles.

Afro-Portuguese carvings were made in Benin (now part of Nigeria), and in Sierra Leone. Many had images of Portuguese gentlemen, including objects made for and used by the Africans themselves. For example, the famous pectoral masks worn by the Oba (or ruler) in Benin, which were said to honour his mother-in-law, depicted a face surrounded by mudfish or small patterns, with a semi-circle of little Portuguese heads around the top. One theory behind this is that the Portuguese were white, and the people of Benin worshipped the god Olokun, who was also white. Olokun brought wealth, children and trade, and the Portuguese brought guns and trade, so there was a similarity and the Portuguese were therefore considered worthy of being honoured. But this is supposition, as the real symbolism is unknown.

The various beliefs and myths upon which African cultures were based are confusing. It has been said that the more one gets to know Africa, the less one understands it. Stories and legends are anecdotal, nothing was written down, and explanations can be very flexible, with the truth sometimes intentionally buried.

The style of ivory carving varies a lot from one African culture or country to another, but it is clear that the styles within the cultures remained unaltered, possibly for centuries. This is because the ivories had definite purposes – symbolic, ritual or talismanic – rather than being purely decorative. It also makes the carvings almost impossible to date with any accuracy.

Certain colours were popular and had definite connotations. The three favourites were red, white and black. Red drove away evil forces, white denoted prosperity, joy and peace, and black was the colour of the people. Some African ivories can have a reddish hue from palm oil. In the Congo and in Ghana mahogany dust was mixed with palm oil as a very effective red stain which could be used on anything, including ivory.

Side-blown horns are possibly the most common form of ivory use in Africa, and were part of many cultures. They came in various sizes, from small ones about 15 cm long, to the massive ones used by the Swahili that could be over 2 m long. For the larger ones, whole tusks were hollowed out and carved, with an aperture and mouthpiece at the side of the tusk and the tone hole at the tip, where it was operated by the user's thumb. Smaller horns were either carved from the tusks of younger animals, or from large tusks that were cut down and carved to look as though they were whole tusks. The different sizes created different notes, and it is said that when three or four were blown in unison they sounded like a human shouting, and that the sound carried a long way. A single horn of standard size is said to sound very like an elephant trumpeting. Horns were often decorated with incised or low relief patterns, as befitted their importance.

They were used in various ways among the different cultures, although usually only for ceremonial purposes. In Cameroon they were blown by the judiciary when settling a dispute. Among the Asante in the Gold Coast human jawbones taken from enemies in battle were attached to the open end of the horns. In many areas the horns were used for passing messages over a long distance.

Another common use for ivory was staff finials. The staffs were used by the chiefs, and the finials that surmounted them were usually figurative. Carved in the round, they usually depicted gentlemen, and great attention was paid to their dress. Most depicted Africans, but a number also depicted Europeans. As is so often the case, the reason for these 'portraits' is unknown, but they seem to have symbolized wealth and status.

All over Africa ivory body adornment was used, though only by high-ranking people. Each culture had its own form of jewellery, and rules about who could wear it and why. Objects worn included hairpins, pendants and beads, and – most commonly – bangles for the arms or legs.

Women in the Igbo culture were adorned with jewellery when they were admitted to a society and given a title. The jewellery consisted of hefty bangles on both wrists and ankles. The latter could be up to 30 cm long, and were so difficult and painful to put on and take off that many women simply kept them on. In the Ivory Coast a man had to give his first, 'senior' wife a large bangle if he took a second wife, and again if he took a third, and so on. He also had to give the senior wife a bangle if he gave one to another of his wives. As a result the first wife could end up with many bangles.

Hair combs and long pins for adornment were also quite common, and the people of the Ivory Coast carved some very elaborate ones. They also made a more unusual item out of ivory: babies' dummies or pacifiers. These were small, round, and mushroom shaped, and were worn by the infant on a cord around its wrist.

In southern Sudan the Nuer used ivory as jewellery. Great significance was given to the correct division of a tusk, which had to be shared among certain eligible people. Failure to do so would cause the spirit of the dead elephant to bring misfortune to those who had portioned out the tusk.

The Kingdom of Kuba, in what is now the Democratic Republic of the Congo, was one of the last areas of Africa to have many very large herds of wild elephants, because of its geographical position. When the interior of the continent was finally penetrated, the local tribespeople found themselves at the hub of various trade routes – a position which they guarded fiercely. For a long time it was possible to see in the region whole ivory tusks firmly planted, tip downwards, in the earth. They had been put there to be used as a backrest for the king when he visited a village, and they were left in place to commemorate his visit.

The Pende and Lega tribes of today's Democratic Republic of the Congo had differing customs, but the males all underwent several initiation ceremonies, progressing in rank with each one. During this time the initiate would be taught morals, philosophies, wisdom and the correct way of life. Each ceremony had a symbolic artefact representing an oral tradition. The object was to be worn or held – they were not for display as trophies. Many were figurative, and they were made of various materials, progressing in accepted value. Wood was lower on the scale, whereas bone and ivory came with the highest ranking. It was the custom to pass the initiation objects down through the generations. As a man progressed, he would pass the relevant object to a younger relative, and so forth.

Among the Pende, some of the initiation symbols were pendants which depicted a head. They were not stained, as the Pendes liked the white colour of ivory, but many today are brown and discoloured, probably from being worn constantly against the skin. The significance of a head as the subject is thought to be connected to the brain and the spirits.

The Lega were not involved in either the ivory or the slave trade. They defended their territory against raids, and much ivory was stolen from them, both raw and carved, but they preserved their way of life and their attitude to both ivory and the elephant.

In Yoruba huge amounts of raw ivory were exported but not so much was used at home. However, the few artefacts made there were of the highest quality, with intricate and technically demanding carving. An example is the broad bracelets made in two interlocking parts, one inside the other, carved in relief and with openwork. These were also made in neighbouring Benin.

As was the custom in many cultures, in the Kingdom of Benin the Oba automatically received one out of every pair of tusks and, if he required the second tusk, he had the right to buy it. He used a lot of ivory, and also stockpiled large quantities.

Benin was renowned for its brass work as well as its ivory carving. The workers had guilds, the members of which had to adhere to very strict standards and practices, overseen by the Oba. The ivory workshops were situated in the palace grounds, and the craftsmen formed an artisan class of their own. They belonged to just a few families and the craft was passed down through the generations. Youngsters had to work out an apprenticeship lasting eight or nine years.

The work carried out in the royal ivory workshops could be divided into two groups: items for the foreigners, and objects for court use. In the first group are the well-known Afro-Portuguese carvings such as the tall salt cellars, consisting of a lidded bowl for the salt on a stand, which was usually decorated around the base with a group of Portuguese soldiers or equestrian figures (see front cover). Other items made for the foreigners were end-blown hunting horns, forks and spoons. There is a lot of detail in the carving and the quality is high.

The second group of objects, those for the Oba's court, were more varied but of equally high quality. Ceremonial apparel included intricately made, wide bangles, and the already mentioned pectoral masks worn by the Oba. In various cultures there was a custom of placing whole elephant tusks on an altar as a tribute to the ancestors, but unique to Benin was the custom of mounting them in the top of life-size brass heads. The tusks were, as with all of the ivory work in Benin, ornately carved.

One unusual use of ivory that came from the Benin royal workshops was in the form of the large leopards, made from six tusks each, which were joined together with ivory pegs. For once it appears possible to date them with some accuracy because the leopards' spots are made from the copper percussion caps from nineteenth-century firearms. They are *aquamaniles* or water carriers, which were usually made out of brass, but because of their construction the ivory ones could never have held water and so were simply for display. The leopard had a certain significance to the Oba, and was connected to beliefs about his ancestry. Following the Punitive War in 1897 the leader of the campaign, Admiral Rawson, brought the leopards back to Britain and gave them to Queen Victoria. Today they are in the British Museum in London.

Over the years the various Obas and their courts had always stayed on good terms with the Europeans who settled in their country: first the Portuguese in the late fifteenth century, followed by the Dutch and the French. As long as good trade continued, there seems to have been mutual respect. Imports into Benin included coral, which was extremely popular at court, cowrie shells which were used as currency, and brass. Firearms were also imported, which meant that elephant hunting could be more successful and bring in more tusks, and more slaves could be rounded up, so there was more trade and more prosperity. Just as important was the promise of support against Benin's enemies.

Terms were less good when the British arrived in the nineteenth century and wanted to annexe the country and change some laws, including not allowing human sacrifice. The British were also determined to ban slavery, which had been outlawed in Britain for almost ninety years. The Oba could see no reason to stop human sacrifices, or slavery. He sold slaves for a good profit, and he had many himself. Following some battles, the British won the Punitive

War in 1897, razed the city of Benin, looted wherever possible, and then annexed the country. Before the campaign the British had calculated the cost, and fully expected to take enough ivory from the Oba's stockpile to finance the whole campaign. The huge quantity of items looted from the country were taken back to Europe and sold at auction.

The story of Benin was being repeated in various forms all over Africa. Countries were being colonized by Europeans, whose own home countries sometimes ceded the African ones to each other in battles of their own. And ivory seems to have been one of the favourite spoils to take back to Europe.

The Kingdom of the Kongo, now called the Democratic Republic of the Congo, had its name changed to the Belgian Congo when King Leopold II of Belgium took it over at the beginning of the twentieth century. By the time it had gained independence in 1960, large numbers of ivory artefacts had found their way to Belgium.

The slave and ivory trades were connected and the values of the two fluctuated, but it is estimated that around the year 1900 one large elephant tusk could be worth the same as several slaves thanks to a rise in ivory prices. In Sierra Leone the Sherbro people were renowned for their trade in raw tusks and slaves. The trade in raw ivory was far more important to them than owning ivory themselves, and consequently very few carved items for their own use have been found. The Sherbro were an important link in the slave trade and became more powerful than the European traders. They did not find the slaves themselves, but employed other Africans to find and capture them.

Because of the huge trade in ivory throughout most of sub-Saharan Africa, many societies hunted their populations of elephants almost to extinction. Not only did they trade bigger and bigger quantities of raw ivory, but they also started to use much more themselves. Their spiritual connection with the animal was being ignored and ivory became more a symbol of wealth and power than one of strength endowed by the elephant. So much was the connection to the elephant lost that in Cameroon in the early twentieth century Germany was trading celluloid bangles for the local population to wear, in return for raw ivory.

The artefacts made from ivory and used by the Africans at this time were much the same as they always had been. Styles had not changed, and with the possible exception of the Swahili who used ivory inlay in furniture, the Africans did not adopt European ways of using the material. The Swahili could, of course, be said to have been influenced by their ancestry, as they were descended from Muslims who had fled the caliphate in the ninth century and married Africans.

The Africans did, however, understand that the Europeans were fascinated by the 'native' styles of things in Africa and were willing to buy them. Thus in

the nineteenth century some cultures began carving ivory items purely for sale to the traders or visitors to the continent.

The Loango people, who occupied the coastal region of what is now the Democratic Republic of the Congo, were known for their whole tusk carvings, depicting everyday life. These were carved in relief, with the decoration spiralling from the proximal end to the tip of the tusk. Images were commonly of scenes of hunting or fishing, and there were many scenes of African porters. There was usually a European figure or two, at leisure, possibly wearing a pith helmet and with a drink in hand. Sometimes animals – usually monkeys – were illustrated, and very occasionally the tusks depicted stories from books.

This form of carving was produced in warthog as well as elephant ivory. Both types were carved in relief, and the quality of the carving varied greatly. The high relief and detailed carving on the best of the elephant tusks is of a very high standard, while the low relief and lack of attention to detail given to the warthog tusks suggests that they were of the 'cheap and cheerful' type of souvenirs.

Warthogs (and related animals of the pig family) and hippopotamuses are plentiful in Africa, and their ivory was also worked, though undoubtedly less than that from elephants. Unfortunately there are few records of how or when the alternative ivories were used. A few spatula-shaped objects

◄ **Fig. 3.24** Whole, small elephant tusk, carved in Loango, probably for sale to foreign visitors. Note the European gentleman (*second segment up*) depicted seated, wearing a pith helmet, and the African carriers. Height 42 cm.

found during excavations in Ghana appear from their shape and curvature to have been made from warthog tusk. It is thought that they date from the seventeenth century, but their purpose is unknown.

Warthogs and hippos feature far less in African art, beliefs or legends than the elephants. Even today there are countless ceremonies and dances which have been passed down through the generations in which elephant masks or costumes are used, but few, if any, symbolizing warthogs or hippos. These animals were not considered to have the mystical and spiritual strengths of elephants, and were not revered in the same way. Furthermore, they had none of the traits that humans like to think of as similar to their own, so they were not endearing animals. For example, hippos are notoriously grumpy, and that trait would not be one that a human would want to absorb, transferred through something made from its ivory teeth.

Fig. 3.25 Warthog tusk, carved in low relief and of inferior quality. Probably carved in Loango. Length along outside curve, 23 cm.

It is known that hippo ivory was shipped to Europe long before the nineteenth century, when it was used for things like cutlery handles. It was also used to make false teeth for wealthy Europeans and Americans as it does not stain as easily as some other ivories, has less of a tendency to crack, and is a little harder. Hippo ivory was, though, never as sought-after as elephant ivory, and, although it was always less expensive, it was not sold in anything like the quantities of elephant ivory. It is, however, impossible to know the exact quantities exported from Africa as old records do not differentiate between the types of ivory. It was only in later years when shipments arrived at destinations such as London, and were sold, that the different forms of ivory were catalogued.

It is likely that there are more hippo ivory carvings in all parts of the world than we are aware of, which have simply not been identified. The material is still carved and sold in some African countries, although importing such souvenirs from Africa would be illegal in most countries nowadays. Today ivory is still used by African peoples, and some may still be carved according to old rites, but it is believed that the amount made is negligible. The artefacts cannot be legally exported.

Fig. 3.26 Small hippo tooth. Modern carving made as a souvenir, decorated with hippos. Note the typical worn area at the end of the tooth.

Since the fifth century BC Africa has had a trade in raw ivory, and the trade grew until Africa was by far the biggest supplier of the material, far outstripping Asia in exports. The trade continued until the trade bans of 1989 came into force. Today it is the centre of attention in regard to the poaching of elephants and the illegal trade of elephant ivory which, as always, is driven by foreign demand and not by the demand for ivory amongst the African people (*see also* Section 1.1 Conservation).

3.7

India and Sri Lanka

India

Early books on Hindu mythology state that Indra, the ruler of the gods, had an elephant named Airâvat, who was said to be the progenitor of the whole elephant race. The Hindu religion has an elephant-headed god called Ganesha.

Elephants have long been part of the culture of the Indian subcontinent. In many areas of the country they were part of everyday life. Tamed elephants were work animals. They were also used in battle, and made up part of the Indian army's 'cavalry'. They also supplied the revered and expensive material: ivory.

Ivory was a precious material, and when it was carved none was wasted. Even the dust resulting from carving was collected and fed to cows or buffalos in the belief that it would increase their milk yield.

In ancient times not much ivory seems to have been used within India. The small amount that has been found in archaeological digs consists of the sort of personal items that are found in many places: hairpins, beads, small vessels, and suchlike.

Yet by the first century AD ivory carving was a respected craft, and records state that in the ancient city of Vidisa (today called Bilsa, south-east of New Delhi) the ivory carvers were organized in a guild. In the ruins of Pompeii (destroyed in AD 79) was found a carved ivory mirror handle which had come from India, verifying the belief that India was exporting ivory – both raw tusks and finished, carved items – to the Roman Empire, along with spices, textiles, and some live animals including monkeys and tigers.

Probably the most important archaeological find of Indian ivory was the so-called Bagram Ivories, found when excavating a palace near Kabul, in

Afghanistan. These date from the first or second century AD and comprise about one thousand ivory and bone plaques, which were once attached to furniture. They depict mostly animals, forms of foliage, and women at various activities (for example, doing their hair, dancing and horse riding), giving rise to the suggestion that they may be pictures of courtesans. The real purpose of the plaques is unknown, but it is clear that they formed part of the luxury lifestyle that was enjoyed at court at the time when they were made.

The plaques are carved in relief and also incised, and were once coloured. Some small remnants of vermillion and indigo pigments remain, and there is speculation that there may have been other, less stable, colours used as well that have since disappeared. There is also black, but this could possibly be an accumulation of dirt. A few of the ivory panels are carved with openwork.

The origin of the ivories remains something of a mystery as the style of carving is mixed, even on a single piece. There is some script which is in an old Indian language, but it is still not known whether the ivories were carved at Bagram by Indian carvers, or whether finished ivories were imported from India. Some of the plaques were looted from the museum in Kabul in the 1970s and were sold, but a number of those purchased were returned as a gift to the museum. They needed extensive restoration as they had not been well cared for (*see* Section 2.7 Working Ivory).

Until the eleventh and twelfth centuries furniture had been virtually unknown on the Indian sub-continent, as everyone sat on the floor, with carpets, cushions and bolsters making those who could afford such items more comfortable. Even then the use of furniture was still very limited and consisted of caskets and boxes for storage, beds which comprised woven wicker bases on short legs, thrones for people of very high rank to elevate them from their subjects, and footstools to go with the thrones. Thrones and footstools incorporated ivory, and sometimes the bed legs were carved in ivory as well.

A little ivory jewellery was used – for example, the customary bangles worn by married women. The bangles were incised with patterns, left plain, or stained in red, green or black. They could be worn several at a time, all the way up the arm from wrist to shoulder.

From the thirteenth to the fifteenth century a little more elephant ivory was used, for instance as inlay on palanquins, door handles, or similar utilitarian items. Ivory was also being exported, but there is no clear evidence to suggest that the valuable elephants were being killed specifically for this purpose.

It was not until the Muslim influence began to be felt in India that ivory was used more extensively for luxury or decorative items. In 1526 Hinduism was pushed aside as the Mughal Empire took control. Influences from the Safavid Empire in Persia (now Iran) were very strong, even to the extent that in Mughal India, Persian became the administrative language of the whole empire.

The Mughals loved precious stones, gold and enamel. They also admired ivory, and over the following years it was used more and more extensively, starting with smaller things such as fly whisk handles, powder horns, and games. Boxes were also made for use for both storage and transport. All these items could be made in other, less costly materials, but ivory had become the material associated with luxury and was only within the economic reach of the wealthy or of royalty, and thus it denoted status.

Not only elephant ivory was used: other types were employed too, notably hippo teeth, while imported walrus tusk was the preferred material for dagger handles as it gave a better grip than elephant ivory, feeling drier to the touch.

In the early sixteenth century the Portuguese, who were expanding their empire and their trade, arrived and made settlements in Cochin and Goa, bringing with them their religion and their customs. One of the first things they required was furniture, as they were not inclined to sit on the floor. They also brought their Christian iconography.

Indian craftsmen were already skilled in making fine things for the Asian market using ivory, tortoiseshell and mother-of-pearl. They were also renowned for being able to copy anything, so furniture was duly produced for the new settlers, the designs being copied from illustrations they had brought with them. The furniture produced had an Indian touch as the woods used were local, and pieces were often inlaid or veneered with ivory. Seeing the wonderful quality of work and use of the precious materials, the Portuguese sent carved ivory caskets back to Europe, where they met with great approval. The new Indian styles were very popular in stately homes and even in churches, and a lot of work was commissioned by the royal courts of Europe.

At the turn of the seventeenth century the British also arrived and the East India Company (or EIC) became established on the Indian subcontinent. The new settlers brought with them more furniture designs to be copied. The furniture was based on Hepplewhite or Sheraton type designs, which the Indian craftsmen covered in a fine veneer of ivory or inlaid with intricate and delicate designs in ivory. The inlay was usually incised to add to the decoration, and the incised areas were coloured with black lac, a pigment made from distilling the lac insect, which was plentiful in India. The colour was permanent and did not rub off.

The Indian people did not adopt the European designs for their own use, but continued with their own preferred styles. During the reign of Shah Jahan (1628–58), a consistent imperial style was used, especially in the ornamentation of palaces. The arabesque, a symmetrical arrangement of scrolling, flowering plants, was the most common design.

▲ **Fig. 3.27** Toilet glass, mirror and drawers. Sandalwood veneered with engraved ivory. Made in Vishakhapatnam, India, in about 1800.

By the mid-seventeenth century some Indian ports had become trade hubs, with items arriving from both Asia and Europe before being sent on to their final destinations. Trade in the ports was lively, there was an exchange of skills and ideas, and the shops and stalls around the ports sold all manner of things. European designers and craftsmen also arrived, and some of them were employed by the Indian courts. Ready-made European-style furniture was available, but the bulk of it was still made to order.

India had its own elephants, yet it is clear that the ivory workers preferred African ivory. A single shipment of ivory arriving in India from Africa could number 3,000 tusks. The reason for this preference was simple and normally had nothing to do with any possible difference in colour or hardness of the material – it was the size. African elephant tusks were considerably larger than those of Asian elephants. They were longer and thicker, and from them could be taken much larger pieces of workable ivory, especially the large, thin sheets needed to veneer furniture.

Although the Dutch East India Company (*Vereenigde Oost-Indische Compagnie*, or VOC) was the largest European presence in Asia, and over the years sent out more than a million people to work in their various colonies and settlements, the company's concentration was mainly on South-East Asia. During the next two hundred years it was the British East India Company that took over most of India, and the Portuguese, Dutch, French and Danish left the settlements and trading posts that they had established. The EIC had their own private army, which they were not afraid to use, and governed the country themselves. When the Indian Rebellion of 1857 took place, the British Crown nationalized the company and took over its possessions and army. India became part of the British Empire, with English as its first language.

With many thousands of Europeans settling in Asia during this time the demand for luxury goods – especially ivory-covered furniture – continued unabated.

Furthermore, in Victorian times, the European fascination with foreign design and products included the goods that were arriving from India. At the Great Exhibition of 1851 in London, Victorian England was able to see what was made in the faraway country of which their monarch was soon to be named empress. Various items were on display, including a throne and footstool which was presented to Queen Victoria by the Rajah of Travancore (now Kerala), which she later called her 'magnificent chair'.

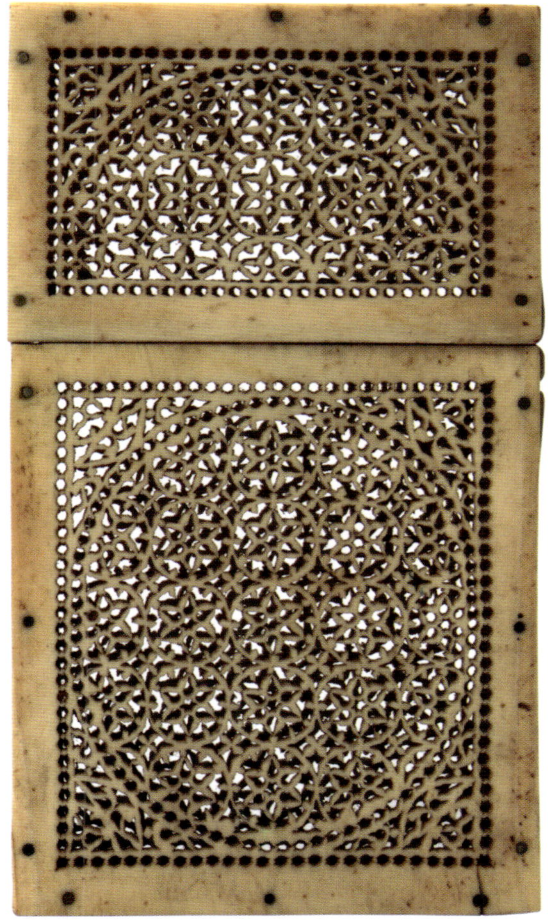

▲ **Fig. 3.28** Pierced work card case, probably made early in the twentieth century.

Very popular with the Europeans were marquetry-covered boxes. These were wooden boxes covered in intricate micro-mosaic marquetry made up of different woods, metals and ivory (*see* Section 2.7 Working Ivory). The method had been developed in Iran many hundreds of years earlier and called *khatamkari* technique. It had spread throughout the Middle East and to India, and by the sixteenth century had become popular there. In India it was called *sadeli* work. The mosaics could be used on any furniture, but were predominately applied to usefully sized boxes. The work is still done today, but not to the same standard, and the ivory has been replaced with white plastic or bone.

Bengal was particularly noted for its ivory craftsmen, but ivory was carved in other areas as well. Although there was some overlap, centres developed in India specializing in the manufacture of certain items. For example, many Hindu shrines were made in Berhampur, and small figures came from the workshops in Marshidabad, both of which were in Bengal, while much furniture was produced in Vizagapatam, in a district of the same name.

The reason for the specialization in different centres was the caste system. All industries amongst the Hindus were governed by a caste, or by a trade guild. For example, the ivory carvers of Murshidabad belonged to the small Bhaskara caste of Vaishnavites.

People kept to their own caste, and castes did not intermarry. The Bhaskars were respectable and earned a decent living when there was work available. Their trade was organized in workshops with master carvers teaching the younger people the craft, which was regarded as just that: a craft, and a way of life, rather than as a business from which they expected to make great profits.

▲ **Fig. 3.29** Two small pots, incised and coloured. Probably made in India at the beginning of the twentieth century.

The carvers were dependent upon dealers for their ivory supplies, as, following the Preservation Act of 1904, buying ivory required a licence. By this stage a lot of ivory was being imported from Africa, but it is said that the carvers of Marshidabad preferred local ivory or that from Burma because of its texture and colour. As they were concentrating on images carved in the round, and not on veneer, they did not need the larger, African tusks, and indeed the more local material may have been less expensive. The finished carved goods were also sold by the dealers, who acted as middlemen. Few of the carvers had their own shops, and they never travelled far from their own homes.

All the work for the Europeans did not prevent the Indian craftsmen from continuing to work for their own royalty and producing fine carvings for their palaces and courts. The Mughal style persisted, and the items made were mostly decorative panels for their palaces or their carriages, although the latter were predominately decorated with silver gilt. Their thrones continued to be overlaid with ivory, incorporating precious gems as well. Any other furniture – such as couches or beds – usually had legs of carved ivory.

The techniques used to work ivory changed little over the years, and were much the same as everywhere else, with perhaps one or two small deviations (*see* Section 2.7 Working Ivory). But while carvers in other countries moved with the times and finally used more sophisticated tools, in India they preferred to stay with the old-fashioned methods. They also stayed with the old styles and motifs.

The ivory inlaid or veneered boxes and furniture produced in India were always of top quality, but the figures and reliefs did not retain their high standard. Features could be somewhat crude in execution, and flowers and leaves could be very stiff in design. Better work was being produced in Japan and China, and in Europe too – indeed, it was even suggested that English craftsmen should go out to India and teach them how to finish the carvings to a higher standard. The Indian craftsmen were producing bigger and bigger quantities of carvings and the quality suffered. The allure of Indian ivory began to wear off in Europe, and in the late nineteenth and early twentieth centuries the craft began to die out and many workshops closed. Had the Indian carvers progressed a little more with the times and also used more modern equipment, they might have been better able to compete with the carvings produced in countries such as China.

Another factor had bearing on the gradual move away from Indian ivory objects. In Europe furniture styles were changing, becoming much heavier and less suited to the delicate work that incorporated ivory, especially the Indian styles of inlay and veneer.

The fall in sales of finished Indian ivory objects in Europe coincided with the local royalty buying less as well. They were no longer as wealthy as before, and they were abandoning their ivory-veneered carriages for motor cars, which did not need ivory decoration.

Sri Lanka (formerly Ceylon)

The influence from India was always very strong in Sri Lanka. When the Europeans settled in the region some European craftsmen went out to work alongside the locals. The resulting work was perhaps a hybrid of many different styles, but it became the country's own style. Much of their ivory carving is considered superior to the Indian work.

Sri Lanka was an independent country, with its own royalty, but other countries were keen to invade. Sri Lanka did not succumb to the colonial rule of the Portuguese and the Dutch, but instead entered into treaties with the invaders. In the mid-sixteenth century a Sinhalese king made a pact with the Portuguese settlers that he would convert to Christianity in return for their protection, thus securing the crown for his successors. To support his promise (which was in fact never kept), he sent several rich ivory caskets, inset with precious stones, to the court in Portugal.

At this time, and during the whole of the 'Kandy Period' (which lasted from the late fifteenth to the nineteenth centuries, when part of Sri Lanka was in foreign hands but the central and eastern areas remained independent), ivory carving was regarded as one of the noblest crafts. The Sinhalese ivory carvers were held in very high esteem in their own country, and enjoyed not only royal patronage, but also the patronage of the settlers, whatever their nationality.

Again there was political unrest, and the British slowly edged the other settlers out, occupying at first only part of the island to ensure that the French did not take it, then occupying all but the Kingdom of Kandy. By 1818 they had taken over the whole island and colonized it. However, respect for the ivory carvers did not cease; indeed, quite the contrary, and the export of carved pieces continued.

The famous Sinhalese caskets were made of solid ivory panels, not of thin panels mounted on wood. The panels were held together with metal strips, pins and hinges, and were very intricately carved in relief. Popular casket shapes were rectangular with domed or angled lids, or circular. The circular caskets were carved from a solid section of tusk, hollowed out, and had a domed lid which was similarly carved from a solid section. Some particularly beautiful round caskets consisted of a hemispherical base and a domed lid, which when closed meant that the item was almost completely spherical. This was an extravagant use of ivory, and the caskets' size was limited by the size of the tusk used. The patterns carved on the exteriors of all the caskets have been described as having 'an exuberance of scrolling foliage', which surrounded dancing figures, hunting scenes, and animals, including elephants. As already stated, there could be a mix of influences, so it was not unusual to find Sinhalese figures surrounded by scrolling vines bearing European-looking flowers. Other carvings made during the Dutch occupation seem to come straight from the Dutch paintings, with motifs such as vases of tulips. Some of the caskets were very ornate, with metal strips set with cabochon cut stones – for example, local sapphires – around the lid, and on the lock.

Fans were another popular item carved in Sri Lanka from ivory. These had a long handle which was beautifully carved, often with openwork, and 'blades' or panels of ivory that opened into a complete or almost complete circle at the top. The blades had ornate profiles but were seldom worked, as the thin sheets of ivory were more efficient at moving the air if they were left plain.

There were some simpler objects made as well, for instance ritual objects for Buddhist ceremonies. These items were usually dyed.

There have been suggestions that Tarshish, the place from where King Solomon obtained the ivory for his throne, along with the apes and peacocks and other exotic trade goods when he ruled Israel in about 930 BC, was in fact the present-day city of Galle in southern Sri Lanka. The idea is possibly a little fanciful as Sri Lankan elephants have a low percentage of tuskers, although three thousand years ago there may have been more. However, among the various locations that have been suggested for Tarshish, Sri Lanka is the only one to which elephants were native at the time.

3.8

China

In China, ivory has either been used to symbolize wealth and high rank, or for trade. Ivory was always a luxury commodity and a valuable source of income.

Elephants once inhabited China, so ivory has always been part of the country's history and it is not surprising that the Chinese have a tradition of carving it that goes back about 3,600 years. An ancient Chinese writing from the sixth century BC states that elephants have tusks which lead to their destruction owing to their use as gifts.

There are very few examples of ancient ivories in China, and knowledge is based on the finds in a few tombs. An example is the tomb of Fu Hao, which dates back to the Shang Dynasty (1600–1027 BC). In it was found the remains of a wooden box containing – amongst other things – two large, ivory-handled beakers and a ewer. They were carved with pictures of dragons, cicadas and geometric designs, typical of the period, and similar to designs found on the bronze objects for which the Shang Dynasty was famous.

The Shang was followed by the Zhou Dynasty. In the ancient city of Qufu, excavation of a tomb revealed ivory plaques, and some personal items such as beads, combs, and thumb rings. Very little else in ivory has been found from the period.

Texts written in the Han Dynasty (206 BC–AD 220) record the use of ivory, but again, very little has been found in burial chambers. However, by this time ivory is known to have been more widely used by those living at the imperial court. It was evidently used less in burials.

The Han started a custom that was to last almost a thousand years, of using elaborate carriages for ceremonial purposes at court. They came in five categories, starting with wood, then leather, followed by ivory, gold and finally jade, which was the most expensive. Only courtly people of very high rank could ride in carriages adorned with ivory, gold or jade. The expensive materials were said to have been used as plaques and finials in the shape of dragons' heads. In some cases the chairs within were also adorned in the precious materials. At certain times people at court also used ivory-adorned palanquins.

By the time of the Tang Dynasty (618–906 AD), there was more contact with other countries, and influences from those countries began to appear. The elephants had disappeared from China, and the Chinese were importing ivory from other countries.

Old records state that ivory was being used as an inlay on furniture in about 1200, but no examples remain. It would not have been a big production as commoners were still forbidden to use precious materials such as ivory, jade, gold and rhino horn. However, it would seem that by the thirteenth and fourteenth centuries the rules regarding ownership of ivory were being somewhat relaxed, as there is evidence of large imports of raw ivory. As not all of it was used at court, it can be presumed that trade in carved ivory objects had begun. Many Buddhist and Taoist figures were produced, which were carved in the round.

During the Ming Dynasty (1368–1644) the Portuguese and Spanish ventured as far as South-East Asia, bringing with them their missionaries and religion, and their desire for Christian devotional items carved in ivory. The Chinese, in turn, were keen to trade with the rich foreigners.

A carving industry developed in the port of Zhangzhou in Fujian province. Working outside the central government or emperor's control, the people of Zhangzhou started a lively trade with the Spanish in the Philippines, producing both carvings of religious figures and ornamental objects. Raw ivory was imported and finished goods were exported from the port. Across the sea lay Taiwan (then called Formosa), which was briefly colonized by the Dutch during the seventeenth century. This was also a new market inhabited by wealthy foreigners.

The Portuguese, too, were active in the area, initially trading legitimately along the Chinese coast, and then for a while involved in smuggling. Finally they settled in Macao, from whence they continued trading, although there is no evidence of Macao having been an ivory carving centre. The Portuguese set up schools and hospitals as well as churches, all of which required carved devotional figures, preferably in ivory. They imported them from India and Europe, and from the new carving centres that were developing in Chinese cities in other areas, including Shanghai, Peking and Canton (the latter two cities now being called Beijing and Guangzhou).

The Europeans had soon discovered that the Chinese were excellent craftsmen, capable of producing very fine work, and also able to copy figures brought from Europe, or pictures in books. Trade thrived, raw ivory was imported from India, Thailand and other surrounding countries, and some began arriving from Africa.

A number of the carvings were coloured with pigments, many of which were toxic (*see* Section 2.7 Working Ivory). The colours adhered poorly to the surfaces and wore off, sometimes leaving a light brown stain where they had been. At this time the tools used for carving were still basic and might include slivers of wet sandstone, and bow drills with drill bits made of moistened bamboo to which grit adhered.

The Buddhist and Taoist faiths influenced the designs used in all carvings in any material, including ivory. Their influence could also be seen in paintings. The Taoists especially had a great love of nature, and hence the depictions of animals, insects, trees and flowers. The various subjects also had a symbolism, for example, the fox symbolized evil spirits, while a fish symbolized wealth.

Unlike in many other countries, the art of ivory carving was not passed down through families from one generation to the next, but practised by individuals who treated is as a job, not as a tradition. Nonetheless, many of the designs that they carved were very traditional, and were passed on from master craftsmen to apprentices.

By the seventeenth century the area of Fujian was becoming prosperous. The carvers and traders realized that many more ivory items could be sold to the domestic population than hitherto, and that it was not just foreigners who wanted them. A huge production of chopsticks, cups, fans and so forth was soon underway, as well as the images of Taoist immortals or Buddhist deities already in production. The figures became a status symbol and proof of wealth, and every home had to have several. In some areas furniture was adorned with little plaques of ivory carved in relief, depicting Chinese rural scenes. The use of ivory suddenly became so widespread that in some circles it began to be regarded as slightly vulgar.

The immortals were not devotional figures, but symbolized various things to which the Chinese aspired, not least longevity. The carved figures were therefore made and purchased for the purpose of providing pleasure. 'Wish fulfilment' was a normal attitude, and figures denoting wealth, happiness, rank and above all longevity, were to be admired. A popular figure was that of Guanyin, a Bodhisattva or 'enlightened being', who was the protector of

▲ **Fig. 3.30** Figure of Guanyin, the protector of mothers and childbirth, with the child Zen Zai. Carved in China (possibly in Zhangzhou), in about 1600.

mothers and childbirth. She was often depicted with a child on her arm, in the same pose as the European Madonna and Child carvings.

Even if the subject matter might appear similar, the Chinese Guanyin looked very different to the Madonna – indeed, Chinese figures had a very different style to their European counterparts. Solitary figures carved in China were often depicted with the arms and the folds of the clothes close to the body, giving an overall impression of a slender, smooth line that followed the curve of the elephant tusk. The folds of the robes were usually curved or rounded at the hem, and the facial features were, of course, Oriental, closer together than in European figures, and had a high forehead.

It has been suggested that Chinese carvers did not understand the significance of carvings of the Christ-child, or cherubs, usually depicted by the Western world as having lots of curly hair. In China babies and small children had their heads shaved, so these figures were probably regarded as images of fat nudes.

The famous 'Doctors' Ladies' were definitely regarded as nudes. For a long time Westerners believed that the figures of the unclothed reclining females were made for the purpose of diagnosing maladies. The idea was that a demure Chinese lady would quietly point to the area on the figure where she herself had problems of some sort, rather than be examined by a doctor. This little story has since been discredited as a fiction dreamed up by prudish Westerners, not least because traditional Chinese medicine seldom requires a patient to disrobe anyhow. These figures were, in all likelihood, erotic carvings, which were popular in the nineteenth century.

In the eighteenth century, under the rule of Emperor Qianlong, carving became sophisticated, intricate, and very beautiful. The Emperor opened workshops at the palace and imported carvers from the Persian Gulf to teach the Chinese craftsmen further skills.

Chinese carving from around this time is considered to be some of the most intricate and delicate made anywhere in the world. Many things were carved

Fig. 3.31 A so-called 'Doctor's Lady', now known to be an erotic sculpture rather than a medical one.

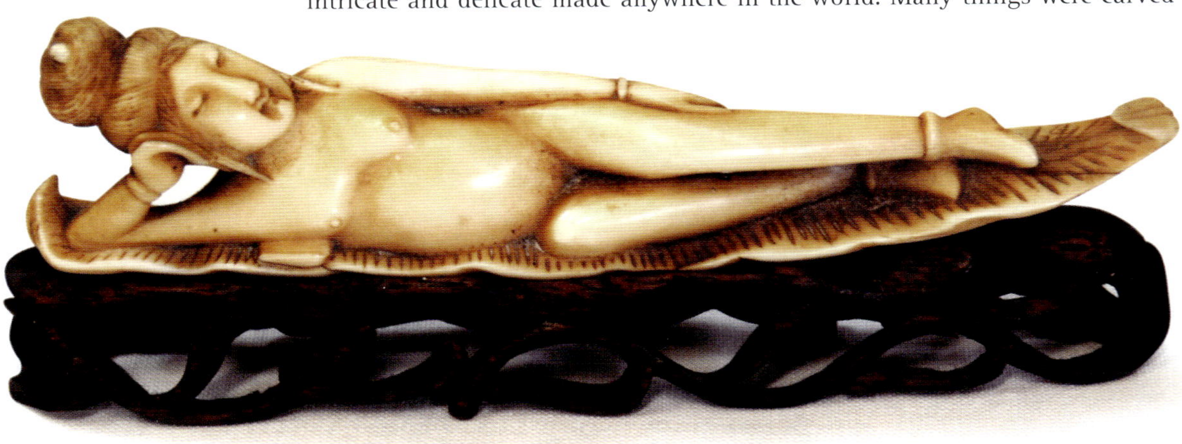

◄ **Fig. 3.32** A puzzle ball, consisting of several balls within balls. Although hand-carved, this example is of the popular and mass-produced variety and not of the finest quality. Height 16 cm.

with exquisitely fine detail, and to complicated designs. One such item for which the Chinese carvers are renowned is the puzzle ball – the balls-within-balls carved from one piece of ivory and not assembled after carving. They were also called devil's work balls, or magic balls. They were made in other countries too – for example, India – but the Chinese were famous for them, making them to stand alone as curiosities, incorporating them in large, multi-layered carvings, or in small versions as decoration on games boards and suchlike (*see* Section 2.8 A Few Worked Items: Puzzle Balls).

Also popular with the European markets were the delicately carved ivory fans and card cases which frequently depicted classical Chinese scenes but also had motifs for Western tastes (for example, the carved area framed in a shape of a cross). More expensive items were the Chinese junks or pagodas with extremely finely carved sections and moving parts, and the whole tusks depicting village scenes.

In the early nineteenth century, Emperor Qianlong's successor, who disliked what he saw as unnecessary luxury and waste, disbanded the carving workshops and little was made for the best part of a hundred years. However, by the early twentieth century the art of carving was enjoying a renaissance, with large amounts of jade being carved, and also much ivory. The buyers were mostly foreigners and the quality of work was high. But yet again carving came to a halt during the Cultural Revolution.

Fig. 3.33 Card case made in China in the late nineteenth century, for export to the West.

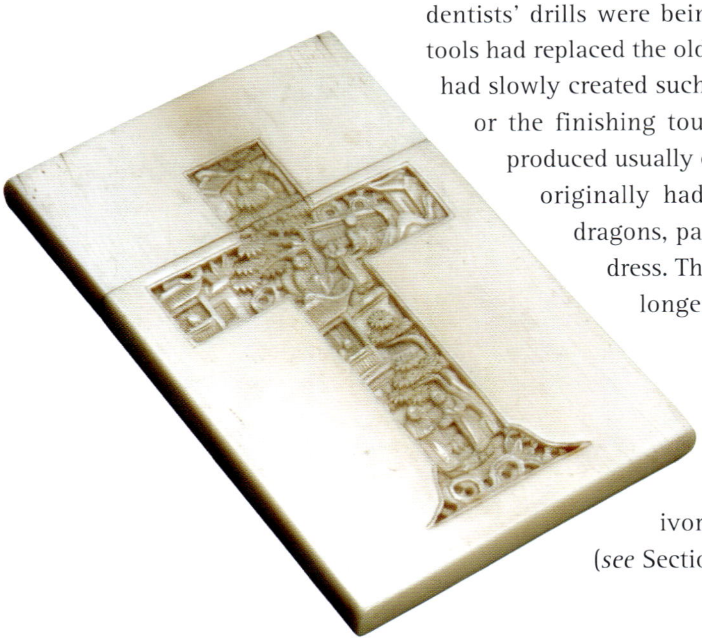

By the late 1970s carving was once again underway, this time with government participation and control, and carried out in large factories rather than in small workshops. Gone were the old tools, and instead high-speed saws and dentists' drills were being used. Diamond and corundum-dust-impregnated tools had replaced the old-fashioned, gentler hand tools with which the carvers had slowly created such delicate and magnificent work. Only the finest work or the finishing touches were made using the old methods. The items produced usually copied old designs with classical motifs, each of which originally had some deeper significance. Typical motifs include dragons, pagodas, bridges, clouds, trees, and ladies in traditional dress. The work was still of extremely high quality, but was no longer as exquisite and individual as before.

Ivory is still being carved in China using elephant ivory obtained legally through CITES controlled auctions, and mammoth ivory which is not affected by trade bans. There are also strong indications of illegal trade in fresh elephant ivory. Only mammoth ivory carvings can be freely exported from the country (*see* Section 1.1 Conservation).

3.9

Japan

Japan's use of ivory is relatively recent when compared with some other countries. A few small artefacts were made in early times, notably plectra for the *shamisen* (a three-stringed musical instrument), seals, and some games pieces. *Bachiru* work was also developed, in which carved pieces of ivory were stained and then incised to reveal the white of the ivory beneath the stain, producing pictures or patterns. *Bachiru* work was also being made in China, and some ivory items were imported from China – a country which exerted great influence over Japan. Yet Japanese homes were minimalist in design, and ivory ornamentation was not part of their tradition.

The best impression of very old Japanese ivory artefacts is to be found in the Shosoin – now a museum – in Nara, Japan. It originated as a series of donations to the Great Buddha of Todai-ji from the Empress Komyo in the middle of the eighth century, and today contains thousands of items from that era, among them some ivory ones. Following that period, however, very little ivory was used in Japan for the next nine hundred or so years.

Kimonos – the preferred attire of gentlemen in the Edo period (1603–1867) and Meiji period (1868–1912) – had no pockets, so small boxes were carried instead, attached to the belt by a toggle called a *netsuke*. The boxes varied in purpose. Some held medicinal herbs, some the very important seals and ink pads (no document was deemed legal unless it had been stamped with the writer's seal), and in later years tobacco and other smoker's paraphernalia. Although each category of box had its own special name, today they all tend to be called *inrō* meaning 'seal basket'.

The boxes were lidded, or could be made in two or three sections. To keep them closed a bead called an *ojimi* was worn. Box and bead were suspended on two cords from the *netsuke*. The items were collectively called *sagemono*, or 'hanging things'.

Sagemono were utilitarian and were often home-made, but it became the custom amongst men of high rank to commission an artist to make them. Over time their utilitarian purposes became blurred and they were worn simply as fashion accessories. The *inrō* were occasionally made of ivory, although lacquer seems to have been the preferred material as it was strong yet much lighter. The *ojimi* and *netsuke* were sometimes delicately patterned in cast metals, but were more commonly carved. Ivory was a favourite material, along with boxwood, horn and amber. In some cases ivory and metal were combined.

Netsuke styles could be dictated by the material from which the item was carved. For example, wild boar tusks were plentiful in south-west Honshu so were an obvious choice of ivory there, but they had to be carved into long, thin *netsuke*, following the shape of the tusk. The carved decoration followed the shape and ran along it, often in the form of leaves, insects or animals. The *netsuke* had to be carved from one piece, as it was not considered fine workmanship if it was made of pieces joined together.

Netsuke were carved in the round and were highly polished, as they were tactile objects. Some were dual purpose, combining their main task as a toggle with a second one as a seal, or as a little box which could be opened to reveal an ashtray, or a tiny flintlock tinderbox lighter for the wearer's pipe. Alternatively they might open to reveal more little carvings inside.

Subjects depicted were mostly taken from the natural world: insects, fishes, frogs, rabbits and other animals, as well as flowers, trees, dragons or demons, and sometimes a combination of these. The twelve animal signs of the Oriental zodiac were popular subjects.

The carvings were in a delicate and flowing style, with soft lines, frequently with the details, such as hair or fur, emphasized by the use of black pigment rubbed into incised lines.

Japan was a closed country for over two hundred and fifty years, allowing no one into or out of the country, and having only minimal trade with Holland and China through one single port, today's Nagasaki. This ended with the death of Emperor Osahito, which also ended the

Fig. 3.34 Netsuke depicting two dragons – one hatching from an egg. Carved in Japan in 1833.

◄ **Fig. 3.35** *Okimono* depicting a samurai warrior on horseback. Walrus ivory. Late nineteenth century. Height 12 cm.

Edo period. It was followed by the Meiji (meaning 'enlightened') period, and Osahito's young son, Mutsuhito, succeeded his father. Emperor Mutsuhito opened up the country so that trade with the Western world could begin again. Artists and designers were encouraged and flourished, showing their work at international exhibitions, where it met with great acclaim.

It would be fair to say that the styles and designs from Japan had an influence on style and design in the Western world, and that they were in part responsible for the Art Nouveau style. Japanese items were in great demand in the West, not least their ivory carvings. *Netsuke* were very popular, and so the Japanese began producing them specifically for that market. Although the Japanese were not themselves accustomed to having objects on display in their own homes, they were quick to learn that other nationalities did so (for example, Victorian Britain). People in many countries in Europe quickly became enamoured with all things Oriental, so the Japanese began making *okimono*, or 'display objects' for export.

There was one other factor that played a role in the sudden production of *okimono*: the wearing of swords in public had been outlawed. Japanese swords were traditionally exquisitely decorated by specialized craftsmen, so many masters in the art of metalwork were unexpectedly without work, and a large number turned to working in other materials, including wood, lacquer, ceramics, textiles and, of course, ivory.

Okimono were larger than *netsuke* and were made almost exclusively in ivory. Elephant ivory was the preferred type, but during the second half of the nineteenth century a huge number of walrus tusks were imported, and many fine *okimono* were made from that.

Westerners liked paintings and carvings to tell a story of some sort, so the custom was carried through to *okimono*, which depicted scenes and people from everyday life in Japan, such as farmers, fishermen, merchants and warriors – all of which were popular with the Western market. As with the *netsuke*, the carving was beautiful and details such as hair and facial expressions were picked out with blackened incised lines. The pigment used was not the usual ink or lac but was a special, secret recipe. It is likely that each family or studio had their own.

It was not long before *okimono* were being produced in great quantity, and the workmanship suffered somewhat as a result. Instead of being carved delicately from a single block of ivory, *okimono* might be carved in many pieces and joined together. This needed less intricate work and was more economical with the ivory, but was considered inferior.

Okimono could be anything from 10–40 cm in height, and the terminology begins to blur with the more complicated carvings. Some extremely fine carvings were produced that were articulated (for example, a crayfish with jointed legs and body). Such items were not generally called *okimono*, even though they were clearly 'display objects'.

In literal translation the Japanese word *shunga* means 'picture of spring', the word 'spring' being a euphemism for sex. *Shunga* art was executed in various media, including ivory. The subjects were not always overtly sexual, but could be highly suggestive. Sometimes *shunga* took the form of a *netsuke* that opened to show tiny figures inside, engaging in some form of sexual act. It was popular in the Edo period but was always frowned upon by the government, which tried to suppress production. Popular opinion said that they were talismanic, but this was probably an excuse for owning them. *Shunga* were also made for foreign trade, and those exported were often far more graphic than would have been appreciated by the Japanese.

Ivory carving was originally carried out by single artists who produced each item individually, taking weeks or even months to complete a single work of art. Many things had to be taken into account, starting with the shape of the piece of ivory, and then the concept of the piece. Even the final process of polishing required skill as the lustre on the surface of a good carving was as important as any other part of its production. This was especially true of *netsuke*, which were tactile pieces and made to be handled. Later it became normal for a master to open a studio where several apprentices worked, taking years to learn the craft before themselves becoming masters. In some cases the art was passed from one generation to another within the same family, but that was not always the case. When work was carried out in a studio it was customary for several people to be involved in each piece, their contribution depending on their level of skill. One person would cut the block of ivory and another start the rough shaping, before passing it on to someone more skilled who would begin to work on some of the detail, and so forth. It was many years before an apprentice was permitted to do any of the fine work.

In Japan it was not considered wrong to copy a piece made by another person; indeed, that was the accepted way that arts and crafts were taught (and still are today). Therefore, copies were not considered forgeries at

Fig. 3.36 *Shunga* or erotic art carving. Possibly carved elsewhere but in a Japanese style. Early twentieth century. Height 3 cm.

the time. It was not until the 1870s, when Oriental art became popular in the West and began to fetch high prices, that forgeries were even considered. Even then a lot of the early 'forgeries' were exquisitely made in their own right.

With regard to the value of a carving, the question is often more about the age of an item than about by whom it was made, and age can be difficult to ascertain. Japanese carvings were usually signed, but signatures can be misleading, as members of the same family belonging to different generations could use at least part of the same signature, and pupils were also allowed to do so if their work passed the approval of their master. It is said that even an old master carver could not always tell whether a carving had been made by himself or was a later copy. The signatures were usually engraved onto the base or back of a carving, but occasionally appeared on an inset red lacquer plaque.

Although styles and subjects remained much the same during the golden age of Japanese carving, there is one style that stood out from the rest, namely Shibayama. In about 1775 a man from a farming family left his home town of Shibayama and went to live in Edo, the capital (now Tokyo). He wished to work as a master carver, not as a farmer. His name was Ōnogi Senzō, and he added 'Shibayama' to his surname. Later his son Shoichi continued and extended the style of work his father had begun, and he kept the name Shibayama.

The technique was to inlay various materials with a variety of other materials, including mother-of-pearl in different colours, tortoiseshell, horn, coconut wood, ivory (sometimes dyed), coral and jade. The background material depended upon the size of the finished object. It was customary to use wood or lacquer for larger items, while ivory was used as the background for small items such as boxes, *inrō*, dishes, letter openers, and games.

In the case of ivory, the background was prepared, shaped and polished, and small cavities created in the surface. Tiny pieces of the materials to be inlaid were cut to match the spaces, and were inserted with fish glue. Mother-of-pearl was almost always used for flowers – white, pink, golden, grey and the iridescent blue/green abalone shell. Finally the piece was polished, softening the edges of the inlaid pieces but not grinding them down to be flush with the surface. They were slightly raised, leaving the design in low relief and giving more life to the finished picture (*see* Fig. 2.42).

Flora and fauna were popular subjects for Shibayama work. Small scenes of sprays of flowers and trailing branches with birds and butterflies are probably the most common decoration on all the ivory Shibayama work. Chrysanthemums (the symbol of the Japanese royal court) were popular flowers, and the Shibayama insects were extremely lifelike. Beautifully precise solitary insects were often used as simple adornment on items such as games boards and pieces, while large objects were decorated with scenes depicting figures, streams, birds, trees and flowers, and outdoor scenes in general.

The Shibayama studio is a classic example of how confusion can arise with regard to the age or maker of any one item. The name Shibayama was used by successive generations of the original family, as well as by many of the craftsmen and followers at their studio. Sometimes it takes one expert to decipher the signatures, and another to sort out carvings.

Netsuke and *okimono* are still being produced. *Netsuke* are especially in demand in the West, and are made in Japan for that market, as well as for their own domestic market where they are now collectors' items. The materials used are mammoth ivory, wood, and tagua nut. For today's modern world, very attractive *netsuke* are also being made in acrylic.

Nowadays China also produces *netsuke* for export to the West, and *Shunga* carvings are also still being made today, largely for the Western market. Many are made in Russia from mammoth or hippo ivory. Care is taken to carve the figures with Oriental features to ensure that they appear to be genuine Japanese or Chinese carvings. They are produced in large numbers and the quality and detail of the carving bears no resemblance to the exquisite carvings produced in Japan in the Edo and early Meiji periods.

3.10

Australasia and South-East Asia

Fig. 3.37 Aboriginal pendant. Wallaby teeth mounted in resin, with red seeds.

Australia

Although Australia is the closest landmass to New Zealand, it has been inhabited by humans for some fifty thousand years or more, compared to only about eight or nine hundred for New Zealand. However, the Australian use of ivory bears similarities to that of New Zealand, mostly because there was little ivory available to the indigenous people, the Aboriginal Australians.

Aboriginal body ornaments usually consisted of shells, brightly coloured feathers and seeds, animal bone and possibly fur, held together with kangaroo sinews, plant fibre, or beeswax. Often they were painted in shades of ochre. Occasionally, wallaby, kangaroo or other teeth were also incorporated.

There have never been mammoths or elephants in Australia, and wild boar were introduced by European settlers in the eighteenth century. There were always whales in the surrounding seas, but, with the exception of the people living in Tasmania or those in the northern parts of the country who went out in canoes and hunted seals, the Aboriginals were not seafaring people and did not venture out to catch ivory-bearing marine animals. Like the Maoris of New Zealand, they used only the ivory that came from beached animals, and, as Australia is a vast continent, this material was only available to a very limited number of people living around the coast. Even there the people tended to prefer shells, so the use of ivory in Australia was negligible. There have been archaeological finds of animal skeletons in middens, especially one in Botany Bay in what is now an outskirt of Sydney, where a few remains of seals, dolphins and whales have been discovered among more numerous remains of shellfish and birds. The midden spans the period from the early years AD to about 1800.

Another archaeological excavation uncovered a skeleton in Victoria that is many thousands of years old, with rows of kangaroo teeth around the head, together with small traces of resin. These are evidently the remains of a head-band with teeth mounted in a resin mixture, which was probably coloured. Another find, in New South Wales and dated about six thousand years old, was a skeleton adorned with a necklace made from one hundred and seventy Tasmanian devil teeth (a carnivorous marsupial now found only in Tasmania).

In the north of the country crocodile teeth were used. These may have had a status connotation as only the best hunter would be able to collect many large crocodile teeth. However, the Aboriginals were not indiscriminate hunters as they were animists and were not prone to killing animals for sport, but only killed them for their own survival.

The arrival of the First Fleet at the end of the eighteenth century, and subsequent British colonization of New South Wales, brought the whaling industry to Australia. Originally the whaling stations were onshore, and the whales were caught with harpoons from small boats. Later, whaling ships were used. The cold waters in the areas around the Bass Strait and Tasmania proved to be good hunting grounds for both whales and seals, and the commercial enterprise focused on oil and skins. There are records of Tasmanian Aboriginal women being forcibly enlisted to help in the hunt for seals because they were excellent swimmers.

Ivory was a by-product of the industry. The Aboriginals had virtually no custom of carving it so there was little interest among them for sperm whale teeth or the small seal teeth. The whalers themselves carved their scrimshaw, and the British settlers imported some finished ivory goods from Europe and China but, although there was a supply of marine ivory, the custom of carving it never became established in Australia.

Whaling stopped completely in Australian waters in 1978 and the following year Australia began to work on whale conservation. This situation has continued in Australia to the present day.

Fig. 3.38 Crocodile tooth pendant. Made for tourists, using farmed crocodiles. Northern Territories, Australia.

Today wallaby and kangaroo teeth are seldom seen made into jewellery, but crocodile teeth from farmed animals make popular tourist souvenirs in Queensland, Northern Territories, and other areas with crocodiles.

New Zealand and the Pacific Islands

About five thousand years ago people set forth from Taiwan and Southern China and headed east towards the South-East Asian Islands, where they settled. Fifteen hundred years later some of them headed further east, inhabiting the Solomon and Bismarck groups of Islands in the Pacific Oceans.

Further explorations took them further east, as far as Fiji, Tonga and Samoa, and the Polynesian Culture emerged. Between AD 500 and 1000 they ventured even further: east to reach the Society, Cook and Marquesas Islands, and north to Hawaii.

Aotearoa (the Maori name for New Zealand) was the last large landmass to be settled by humans. It was long believed that the Polynesian explorers reached New Zealand about one thousand years ago, but more recent research suggests that it was not until the twelfth or possibly even the early thirteenth century that they did so. New Zealand is far away and very remote on the globe, so it is possibly not surprising that it took so long to be discovered by dug-out canoe. Its closest neighbour is Australia, over 2,500 km away. The settlers in New Zealand became known as Maoris.

As all these journeys were undertaken in small boats, the explorers were able to take little with them in the way of goods, although it seems that they did take their domesticated dogs and chickens. They also took their cultures and languages. It is therefore unsurprising that the people from the various islands had similar values, and that their beliefs, customs, and their use of ivory bore some resemblance too.

NEW ZEALAND

The Maoris were animists, as were the people of the Pacific Islands. They killed animals only to eat, and they did not waste any part of the animal. Many animals that they encountered had teeth, but few had teeth large enough to be worked and thus called ivory. Dolphins and seals were abundant, but even their teeth are relatively small. The only animals with large teeth that could be worked as ivory were sperm whales and orcas, but both species are very big and extremely difficult and dangerous to try to catch from a dug-out canoe. However, animals die naturally, and occasionally they also beach themselves.

A beached whale would have been a gift to the Maoris, especially a sperm whale as it contained huge amounts of meat, oil for burning, and teeth that could be carved into utilitarian or talismanic items. Being animists, the Maoris believed it necessary to hold a ceremony to release the soul of the dead animal before cutting it up – indeed, this is a custom that is still observed today by those who retain their beliefs.

Maori carvings were created in many different materials, including the local nephrite jade, shell and ivory. They always had a purpose or a symbolic meaning, and no carving was made purely for adornment. Some items were made for everyday use (for example, fish hooks and harpoon points). These were occasionally made from ivory, but the majority were carved in shell as it was more effective and far more easily obtainable.

Fig. 3.39 Modern *hei-tiki* figure pendant made from bone, with a paua shell eye.

The earliest known use of ivory by the Maoris dates back to the fourteenth century, and consists of small ornaments carved from whale teeth. There is no reason not to believe that they started carving them from the moment they arrived in Aotearoa one or two hundred years earlier.

Possibly the best known of the Maori carvings are the pendants that are worn on a leather thong around the neck. There are various basic designs, representing human forms, fishes, dolphins, and so forth, as well as the famous *hei-tiki* or totem figures. All the various designs symbolize characteristics or emotions, such as strength and friendship. Ivory *hei-tiki* usually have eyes inlaid with the local paua shell, which has a blue-green hue.

Today the pendants are very popular among tourists visiting New Zealand. Some are inexpensive and mass-produced, but there are also very fine carvings made by Maoris who have carried on the tradition. The styles have become more decorative than symbolic, and ivory is no longer used, but has been replaced by bone.

The Dutchman Abel Tasman was the first European to discover New Zealand in 1642, but he was repulsed by the Maoris and was unsuccessful in his attempts to land and colonize the place, so it was not until 1769 that Captain Cook arrived and, after an initial skirmish, made friends with the native population and claimed New Zealand as British.

With the advent of the British came whaling fleets, catching sperm whales which were rendered to make oil, and using sperm whale teeth to make scrimshaw and such items as snuff mulls and powder horns. The new settlers brought a meat source with them in the shape of boar, which they released in the forests, where they have thrived ever since. Some things were made locally from boars' tusks, but being small and largely hollow the tusks were of limited value as ivory. When used they tended to be left whole and undecorated, and simply polished. They were made into a variety of items, ranging from toast racks to pendants, and also handles. The settlers brought with them their custom of using ivory for handles on canes, parasols and teapots, although because of the relative lack of native ivory such items were mostly imported ready-made (*see* Fig. 1.54).

THE PACIFIC ISLANDS

The custom of using teeth as personal decoration is a worldwide phenomenon. The countries which had no animals with teeth large enough to be worked and called 'ivory' were no exception. Small teeth that were available were drilled and strung on a thong made of leather, cloth or plant fibre, or of plaited human hair.

In many of the Pacific Islands porpoise teeth were made into ear adornments or necklaces, frequently with coloured beads added. The teeth were small, and rather than make delicate little necklaces of single rows of teeth, they tended to be strung in chunky ropes of numerous teeth. Headdresses, bracelets and breastplates were also adorned with various materials including ivory, the last often derived from wild boar. Other teeth utilized included dogs' teeth, bats' teeth and even human teeth. Sometimes the items of adornment had additional symbolism and could signify status, or might be used as currency.

Fig. 3.40 Breast ornament of boar tusk with pendant shells and abus beans. Humboldt Bay, West Papua.

Whale teeth were large enough to be worked, but nonetheless were often left plain and simply polished. They formed part of ceremonial dress and were predominantly worn by men.

There are many examples from Fiji of a row of whole or split whale teeth strung together and worn around the neck, possibly as a show of status, but in Fiji a single whale's tooth was also an important cultural item. It was known as a *tabua* and was often given as a gift. *Tabua* were of sufficient importance for men to be buried with them, as well as with their weapons. Inevitably *tabua* became attractive as tourist souvenirs, but now ivory ones are only sold under strict licence and with the appropriate documentation, and have been largely replaced by copies and fakes.

In Fiji and in other areas such as the Solomon Islands whole whale teeth were used as staff heads or handles – the staff itself was made of some form of wood. It appears that the islanders also knew how to soften the thin, proximal end of the tooth which had been in the jaw, as this part often appears to have been crimped to fit neatly onto a cane or similar.

Not all ivory was left plain: in the Marquesas Islands ivory (though more usually bone), was carved into small items such as pointed ear piercers, and plugs to put through the hole afterwards. The ear-plugs were often decorated with little carved *tiki* – the totem figures used throughout the Pacific.

Each of the islands, or groups of islands, made beads and carved them with their own favourite motifs, all of which had symbolic meaning. They varied from place to place: in Tonga a popular subject was a female figure which was a fertility charm, while in the Marquesas and Cook Islands they carved stylized images of testicles. In many places they carved their own style of *tiki*.

In Hawaii another ceremonial pendant could be found: the *lei niho palaoa*. These were hook-shaped and carved from sperm whale teeth. They hung on thick braids made up of numerous thin braids of human hair. The source whale was not hunted, and only teeth from beached whales could be used. Furthermore, the whale was regarded as being the property of the king. However, it was permitted for the flesh to be eaten and the teeth carved. Apart from *lei niho palaoa*, the teeth were used to make inlay, or carved to make other small ornaments. *Lei niho palaoa* were not dyed, but were left in their natural, pale colour, while other ivory objects might be darkened. This was done by wrapping them in the leaves of a local plant, and smoking them over a fire.

Lei niho palaoa denoted status and were originally worn only by people of very high rank. They could be worn by either men or women. When whalers brought in walrus tusks in the nineteenth century, ivory became a less rare material. The pendants were carved in the new material and became more widely used. They also became larger, and the braided hair was correspondingly thicker. It has been suggested that the larger walrus ivory pendants were in fact made for sale to Europeans and Americans, and that they had no meaning to the islanders as they were not connected to their local whales.

The advent of the whaling fleets altered the ivory situation in the Pacific Islands. Suddenly it was possible to obtain plenty of sperm whale teeth, and they were used as barter by the whalers who exchanged them for ships' supplies such as coconut oil. Meanwhile walrus ivory was becoming less valuable in Europe, so the whalers also traded Pacific walrus tusks with the islanders, who were delighted to be able to get so much of the beautiful material to carve. They made the pendants mentioned and other small objects in native designs, and sold them to foreign visitors. They also sold the finished items back to visiting whalers.

▲ **Fig. 3.41** Trading necklaces from Papua New Guinea. Dogs' teeth (top) and porcelain teeth (bottom).

There is little indication that whale ivory was much used in Papua New Guinea, though undoubtedly some would have been used when it was available, just as it was everywhere. However, there was a custom of using dogs' teeth. The people of New Guinea have always been traders, each area having its own trade ring, where items were continually passed around and around – sometimes over many years and large distances – in trade and barter. The favourite trade item was shells, but dogs' teeth were also considered very valuable. Indeed they were so highly valued that during the German colonization at the end of the nineteenth century, the German settlers sent real dogs' teeth back to Germany to be copied in porcelain, and the fakes were subsequently introduced by them into the trading system, where they were regarded as special 'white man's dogs' teeth' and valued as highly as the real ones. The teeth, both real and porcelain fakes, were bound together with coloured cloth, though it seems that they were kept in separate strands and not mixed together.

SOUTH-EAST ASIA

Only the male elephants of South-East Asia bear large tusks – the females have small tushes. Even so, there was no lack of ivory in the South-East Asian countries. Ivory was held in high esteem, but the elephants themselves were also valuable as beasts of burden and working animals. Until the nineteenth century they were used in battle too. Therefore (although it is known that some raw ivory was being sent to China, possibly as far back as about AD 200), in times past the region did not see the wholesale slaughter of elephants for their ivory that occurred later.

In Thailand excavations have uncovered only a few examples of old ivory. One example consists of some ivory bead necklaces found in a grave dating from about 3600 BC. Later graves disclosed combs and bangles from about 500 BC.

Even as late as the end of the nineteenth century, the ivory used in Thailand came from mostly from elephants that had died of natural causes. It was used to make ceremonial artefacts to be used by royalty, and into small items such as door knobs or carvings of the Buddha.

When demand for ivory grew, with more needed for export to China and later for the Europeans as well, the male elephants in Thailand also began to face slaughter. Elephants with tusks became valuable as a source of ivory, and those without tusks could be put to work. Nevertheless it was not until the early twentieth century, when the Thai royal family started to commission a few ceremonial items, that the use of ivory began to increase within the country itself.

Ivory carving really began in earnest in the mid-twentieth century when a Buddhist monk decided to use some of the ivory stockpiles from his temple to carve figures such as elephants or Buddhas. They were to be sold to visitors in order to raise money for the temple and the school attached. With the advent of tourists in the 1970s, demand for these beautiful little souvenirs grew, and soon several monks were learning the craft and selling the objects. Ivory carving became so important that it was adopted as a subject taught in the Fine Arts Schools.

Within a few years demand outstripped supply. Mass-produced ivory jewellery was being made locally but could still not meet the demand, so the shortfall was imported from China. Naturally the quality of the mass-produced items was not of the same standard as before.

The trade bans at the end of the twentieth century put a stop to the sale of ivory trinkets to tourists. The schools ceased to teach the craft and many ivory carvers lost their jobs, but the industry continued underground. It is reckoned that in the middle of the eighteenth century there were almost a quarter of a million elephants in the country, but now there are at most six thousand. Today Thailand is one of the hubs in the illegal raw ivory trade, which the Thai government is endeavouring to stop.

There has always been a custom of producing fine carvings in Myanmar where they worked in wood, jade, and other materials, but, like the Thais, the Burmese did not have a custom of carving ivory. However, it is thought that they exported raw ivory to countries under Portuguese occupation where ivory was in demand by the local carvers, who were keen to supply religious carvings for the new settlers.

The Burmese attitude to ivory changed when the country came under British occupation in 1824. The British wished to purchase useful, small ivory goods such as combs or pots. A few carvers started workshops in the main cities, and from then on the craft was passed down through families. By the middle of the nineteenth century, interest in ivory carving had taken off further in Burma, with a few items being commissioned by the royal court. They were mostly small, but in 1878 King Thibaw commissioned an ivory throne. The ivory probably came from animals that had met natural deaths, or from stockpiles in the temples where the Buddhist monks had collected tusks both from their own domesticated elephants and from worshippers who had offered them as gifts.

Surprisingly, this interest did not die out during the Second World War, as the Japanese invaders were keen to buy ivory items such as chopsticks. In the 1970s and 1980s ivory carving reached its peak when mass-tourism arrived in the country. The art waned in the 1990s, but, as Burma was not itself a signatory to CITES at that time, ivory was again exported raw to countries willing to ignore the trade bans.

Burma can also lay claim to the production of an unusual ivory item: ivory hip prostheses. The first ivory prostheses were made for a German surgeon by the name of Themistocles Glück in the 1890s. He experimented with various ivory joints, including hip and wrist joints. However, it was a Burmese surgeon, Dr San Baw (who was chief orthopaedic surgeon at Mandalay General Hospital), who perfected the use of ivory hip prostheses and used them in over three hundred hip replacements that he performed between 1960 and about 1980, in spite of the fact that metal prostheses were available from 1940. It is said that the success rate with ivory was very high.

Today there is still a small local trade in ivory carving in Burma, which is legal if the tusks come from old stockpiles or from animals that have died naturally. Indeed, the carvers are mostly devout Buddhists, who would not wish to carve ivory from an animal that had been slaughtered. They sometimes make offerings in the temples for the reincarnation of an elephant whose ivory they have used. But the demand for both carved and raw ivory is still a temptation, and the less scrupulous carvers may attempt to 'age' a carving by steeping it in tea, coffee, or tobacco juice, in order to make it look old and therefore legal for import to other countries that follow the bans more stringently.

Most of the South-East Asian countries have similar histories concerning ivory. Their own elephants have been hunted almost to extinction – for example, it is reckoned that there are fewer than one thousand left in Laos, while Vietnam's wild elephants can be counted in tens. Most of the countries have a long history of carving various materials, especially wood, and Indonesia is renowned for its beautiful carved work. But none of the countries had a long history of ivory carving.

The ivory industry grew up around the wishes of the people who settled or traded in the South-East Asian countries, and today the illegal trade is fuelled not from within their own borders, but by demand from other countries.

The ivory situation in the Philippines is different from that of the other countries of South-East Asia, because elephants were not native to the Philippines. Yet it became an ivory carving centre when it was colonized by the Spanish in 1565.

The Spanish brought with them their religion and built churches and monasteries. It was their custom to furnish these with ecclesiastical items made of ivory. The Filipinos proved to be excellent carvers, and they were soon carving numerous types of religious figures for their new Christian rulers. A favourite figure was of the Christ-child as the Good Shepherd, which was depicted literally with a figure of a child surrounded by sheep.

As the years passed and ivory was no longer in the sole domain of the Church, the craftsmen turned their hands to all manner of other ivory goods as requested by their European customers. Finished goods were sold to the local settlers, and also sent back to Europe.

The tradition has continued to this day, and there is still a belief among many of the local people that the Christian God can only be truly glorified by the most valuable, pure and special iconography, so carvings have to be made of ivory. Also, the Philippines is one of the main hubs for the illegal trade in raw ivory being transported from Africa to China. The Filipino government is not complicit in this and publicly destroys any tusks that it confiscates.

3.11

The Americas and the Caribbean

► **Fig. 3.42**
Modern scrimshaw on mammoth tusk, by Jane Tukarski. Note that the 'bark' or outer surface of the tusk has been left intact to frame the illustration.

North America

North America did not have a great tradition of carving ivory. The only animals with teeth or tusks large enough to be of use to ivory carvers were wild boar (sometimes locally called razorbacks), and their tusks were of limited value because of their size, curvature and the fact that they are largely hollow.

Before the War of Independence, ivory items were brought over from Britain for the personal use of the British people living there, but little ivory was used locally.

As with other regions where ivory carving was not a custom, animal teeth were sometimes employed. Strung as a necklace they denoted hunters of stature, and were used as status symbols by some Native Americans. Beaver teeth were good for utilitarian purposes as well, as they had extremely sharp cutting edges. Bear teeth are said to have made excellent fish lures.

It was not until raw ivory tusks were imported and turned into luxury items that ivory became part of the American way of life. Some utilitarian items were produced in the carving factories, such as combs, brush backs and small pots for cosmetics, but there were two main uses for raw ivory that caused it to be imported in such huge quantities that it impacted on the elephant populations in Africa.

Around the turn of the twentieth century, huge amounts of ivory were being imported into the United States to be made into billiard balls and piano keys (*see also* Section 2.8 A Few Worked Items). To have music in the home usually meant producing it yourself, and the most popular instrument was the piano.

So great was the production of piano keys in a town in Connecticut that it was renamed Ivoryton.

As in Europe, the game of billiards – and other associated games – became very popular in North America and billiard balls were produced in large quantities. These balls had been made from different materials, including wood, but by far the most successful was elephant ivory.

A third major use of ivory in the United States was, of course, scrimshaw, the art of the whaler. Scrimshaw was made by whalers of all nationalities right around the globe, but it is often regarded as an American craft. There were many whaling ships sailing from the east coast of the United States, and today many of the old ports have museums dedicated to the history of whaling, each with a good collection of scrimshaw artefacts.

The term 'scrimshaw' was used to denote any work executed by the whalers on various parts of the animals that they caught, so that handles or set squares and so forth which were made from the whales' bones were also scrimshaw. Umbrella spokes were made from baleen (from the whale species that have baleen plates in their mouths instead of teeth), and ladies' corsets were stiffened with baleen stays. Some of these were 'scrimmed' or etched in the same way as whale teeth. Ivory was used mostly for special items that could be either utilitarian or purely decorative, and sometimes a mixture of both. Today the term 'scrimshaw' most commonly refers to the etched and decorated whale teeth or walrus tusks that were made solely for display or to the modern scrimshaw which is generally made in bone or mammoth ivory (*see also* Section 2.8 A Few Worked Items).

Scrimshaw became so popular and valuable that a lot of fake scrimshaw has found its way onto the market, some of which is quite difficult to identify, especially if it is mounted. In America it has been given the name: 'fakeshaw'.

Today, the laws in the United States governing the use, import and export of any ivory goods have been greatly tightened, but mammoth ivory can still be carved and sold, as can fossilized walrus ivory. Most of the carvings sold are very simple, such as cabochon earrings, bangles, fossilized walrus tooth slices made into necklaces, or composites incorporating poor quality mammoth ivory (*see* Fig. 1.3, mammoth ivory cabochons, and Fig. 1.9, mammoth slice pendant). There are a few carvers specializing in large and intricately carved pieces, mostly in mammoth ivory. Both materials are usually obtained from Alaska.

Fig. 3.43 Fake scrimshaw sperm whale tooth, made of synthetic resin and cast in a very fine, flexible mould.

Central and South America, and the Caribbean

Pre-Columbian cultures in Central and South America had a long history of carving bone, and the earliest finds date back to somewhere between 10,000 and 8,000 BC . However, there are no indications that ivory was used.

Mammoth and mastodon bones have been found in Mexico and in the Andes, and it is thought that all remotely usable parts of those animals would have been utilized. It is entirely possible, therefore, that their tusks were also used, although no concrete evidence remains. Necklaces made from teeth were used, notably shamanistic necklaces using jaguar teeth and skins.

Nonetheless, there are examples in existence of very fine carving work in bone: some incised, some carved in low relief, and a few examples carved in the round. The preferred bones were those from animals such as deer and jaguars in Central America, and llamas further south. Carved and ornamented human bones have also been found.

Bone carvings with geometric patterns, dating from about 800 BC, have been found in Peru. Bone figures carved by the Incas in the eleventh century have also been found, as have bone carvings left by the Aztecs in Central America, dating from the fourteenth century.

The Mayan Culture was one of the most advanced, and most civilized, in the region. They also carved bone; their carving styles evolving through the centuries. However, it was not until the Spanish and Portuguese Conquistadors arrived that ivory began to feature in the area.

No ivory-bearing animals were hunted in the regions in the sixteenth century. The new settlers brought some animals with them, and it is thought that they may have introduced both domesticated pigs and tusk-bearing wild boar to South America. They also brought their religion, as they did wherever they went, and their need for carved ivory religious icons. However, the settlers traded far and wide across the oceans, and had access to ivory through trade. It is most likely that what they imported to South America was in finished, worked form, and not raw tusks or teeth.

There is even less evidence of ivory being used in the Caribbean Islands than in South America. There were no indigenous ivory-bearing animals on the islands, though, again, pigs and wild boar were probably introduced by the European settlers.

As with other areas around the globe where it was not customary to work ivory, the local peoples used animal teeth as jewellery and adornment, although among those with access to the sea, teeth were usually less popular than shells.

3.12

The Himalayan Countries

The Himalayan countries used very little ivory, although in some of them elephants were plentiful.

It would seem that, in a way similar to the countries of South-East Asia, in the Himalayan region it was not felt that there was a need to kill elephants in order to use their ivory, and only when an elephant died of natural causes were the tusks used.

Two other factors probably helped to preserve the elephants. First, the animals were revered and held in high esteem, and were important in both the Buddhist and Hindu faiths. Second, the countries were sufficiently inaccessible to the rest of the world for the elephants to be at risk of being killed off by poachers.

Nevertheless, there are some instances of ivory use. In Nepal there was a strong influence from the Indian subcontinent, and among the ivory items that have been found some may in fact originate from there. Both the style of carving and the utilization of ivory in Nepal are similar to those of its neighbours before colonization.

A few examples of carved ivory combs dating from the eighteenth century have been found in Nepal, along with mirror backs and small plaques that were probably used as finials on furniture. Larger items include throne legs carved from solid ivory, and Hindu figures. Ivory sword and dagger handles have also been found.

Today bone is carved, mostly for the tourist market. It is often treated to make it appear old.

Tibet had no native elephants, and the Tibetans are much better known for their use of bone than for their use of ivory. All types of available bone were

◄ **Fig. 3.44** Pair of throne legs, carved from solid ivory. Made in Nepal in the twelfth century.

used, including human bone. The material was carved in low relief and, since the Tibetans have a love of colour, many carvings were inlaid with gem materials and might also be painted and gilded. Prayer beads were made from bone – though never human bone – and were usually left unadorned.

To the Western world the idea of turning the top of a human skull into a bowl is macabre, but such objects were reserved for ceremonial use and were not for everyday utilitarian use. The fact that this was once a custom has been used as a sales gimmick for selling so-called 'Tibetan skull bone beads' which can be purchased today. These have nothing to do with Tibet and are simply ox bone beads carved in the shape of skulls.

Any ivory used in Tibet had to be imported. Chunky, ivory discs with a hole in the centre were threaded onto plaited hair as adornment, mostly by men. They could also be removed and used for meditation, somewhat like prayer beads. Tibetan hair rings can be found for sale today, but they are often faked and made from two pieces of bone slotted one inside the other, polished into a smooth shape, and cracked and dyed to give an aged look. They display the recognizable structure of bone. Ivory has also been used to make little boxes or reliquaries, as well as plaques and other small items such as personal seals – a custom probably borrowed from the Chinese.

Fig. 3.45 Two young Asian elephants.

Bhutan has a small population of wild elephants, which migrate to grasslands in India in the summer months, returning to the forests of Bhutan in the winter. As with Nepal, a few ivory artefacts have been found in Bhutan, but again it is not possible to be certain that they were carved in Bhutan rather than in India.

Appendix I

Main Identifying Features of the Different Ivories

Mammoth

TUSKS

Whole tusk:

- Very large, with sweeping curve.
- Dark surface, smooth or rough.

Cross-section:

- Round.
- Narrow cementum layer may be visible at edge.
- Possible cracks at edge.
- Dentine with intersecting arc pattern (Schreger Lines). Angle less than 90 degrees.
- Dark dot at centre, or hole.
- Possible concentric growth rings along Owen Lines.
- Possible concentric cracks along Owen Lines.

Longitudinal section:

- Irregular parallel lines.

MOLARS

- Ridges across surface ('running shoe pattern').

Cross-section:

- bands of enamel and dentine.

Mastodon

TUSKS

Whole tusk:

- Smaller than mammoth, less curved.
- Dark surface, smooth or rough.

Cross-section:

- Round.
- Narrow cementum layer may be visible at edge.
- Possible cracks at edge.
- Dentine with intersecting arc pattern (Schreger Lines).
- Angle more than 90 degrees, like elephant ivory, but slightly finer structure.
- Dot at centre, or hole.

- Possible concentric growth rings along Owen Lines.
- Possible concentric cracks along Owen Lines.

MOLARS

- Cone-like cusps on surface.

Cross-section:

- similar to human molars, not bands of enamel and dentine.

Elephant

TUSKS

Whole tusk:

- Large, curved.
- Pale, smooth surface.

Longitudinal section:

- Irregular parallel lines.

Cross-section:

- Round to off-round.
- Narrow cementum layer.
- Dentine with intersecting arc pattern (Schreger Lines), at more than 90 degrees.
- Possible concentric growth rings in dentine.
- Dot at centre, or hole.

MOLARS

- Ridges across top surface (running shoe pattern):
    - If lozenge shaped = African elephant.
    - If elongated oval shaped = Asian elephant.

Cross-section:

- bands of enamel and dentine.

Walrus

TUSKS

Whole tusk:

- Very solid, with slight curve.
- Smooth surface, irregular shape at base.
- Enamel tip, possible wear through to secondary dentine along front surface.
- Longitudinal cracks.

Cross-section:

- Irregular oval shape with some indentations.
- Primary dentine with faint concentric lines.
- Secondary dentine with globular structure.
- Clear demarcation between primary and secondary dentine.

TEETH

Whole tooth:

- Peg-shaped.

Cross-section:

- Round to oval shape.
- Thick cementum layer.
- Dentine with possible dark area or trace of globular secondary dentine at centre.

Hippopotamus

CANINES, UPPER

Whole tusk:

- Slightly curved, with tip worn away at an angle.
- Longitudinally ridged.

Cross section:

- Kidney-shaped with indentations.
- Curved TIZ.
- Demarcation line between cementum and dentine.
- Very fine concentric lines in dentine.

CANINES, LOWER

Whole tusk:

- Large, strongly curved, with tip worn away at an angle.
- Longitudinally ridged.

Cross section:

- Vaguely triangular, with indentations.
- Curved TIZ.
- Demarcation between cementum and dentine.
- Very fine concentric lines in dentine.

INCISORS

Whole tooth:

- Peg-shaped, may be long (especially bottom middle incisors) but can also be blunted.
- Slightly curved, or straight.

Cross section:

- Round or slightly triangular.
- Very fine concentric lines in dentine.
- Vague dark dot at centre.
- Possible radial cracks.

Narwhal

TUSK

Whole tusk:

- Long, narrow, with anticlockwise spiral and smooth tip.
- Ridges following spiral.

Longitudinal section:

- Indication of spiralling structure, even when polished.

Cross-section:

- Round, with fine indentations and possible fine cracks reaching into dentine.
- Hollow centre, or dark spot near tip of tusk.
- Cementum layer clearly visible. Slightly darker with concentric lines.
- Dentine paler, with fine, faint concentric lines and possible discolouration following cracks.
- Clear demarcation between cementum and dentine.

Sperm Whale and Orca

TEETH

Whole teeth:

- Peg-shaped , or wide at base if from a young animal.
- Blunted tip of tooth, or ring near tip indicating lack of cementum = sperm whale.

Longitudinal section:

- Thick cementum with fine lines.
- Dentine with visible cone-in-cone structure.
- Clear demarcation between cementum and dentine.
- Cementum not reaching to tip of tooth = sperm whale.
- Dark line from pulp cavity to tip of tooth.

Cross-section:

- Concentric lines in cementum.
- Fine concentric lines in dentine.
- Hollow centre, or dark spot at centre.

Wild Boar

TUSKS

Whole tusk:

- Curved and pointed.
- Enamel on two surfaces.

Cross-section:

- Triangular or vaguely oval.
- Hollow for much of length.
- Faint, concentric lines in dentine.

Cross-section, upper canines:

- Slightly dumbbell shape.
- Solid for much of length.
- Irregularly spaced concentric lines in dentine.
- TIZ, usually straight, or hollow.

Cross-section, lower canines:

- Triangular shape.
- Hollow for much of length.
- Irregularly spaced concentric lines.

Warthog

TUSKS

Whole tusk:

- Curved and pointed, especially upper (larger) canines.
- Possible enamel tip.

Barbirusa

TUSKS

Whole tusk:

- Upper tusk: curved, can be almost full circle, point not worn.

Cross-section:

- Upper tusk: round, hollow for much of length.

Appendix II

Dichotomous Key of Different Ivories and Imitations

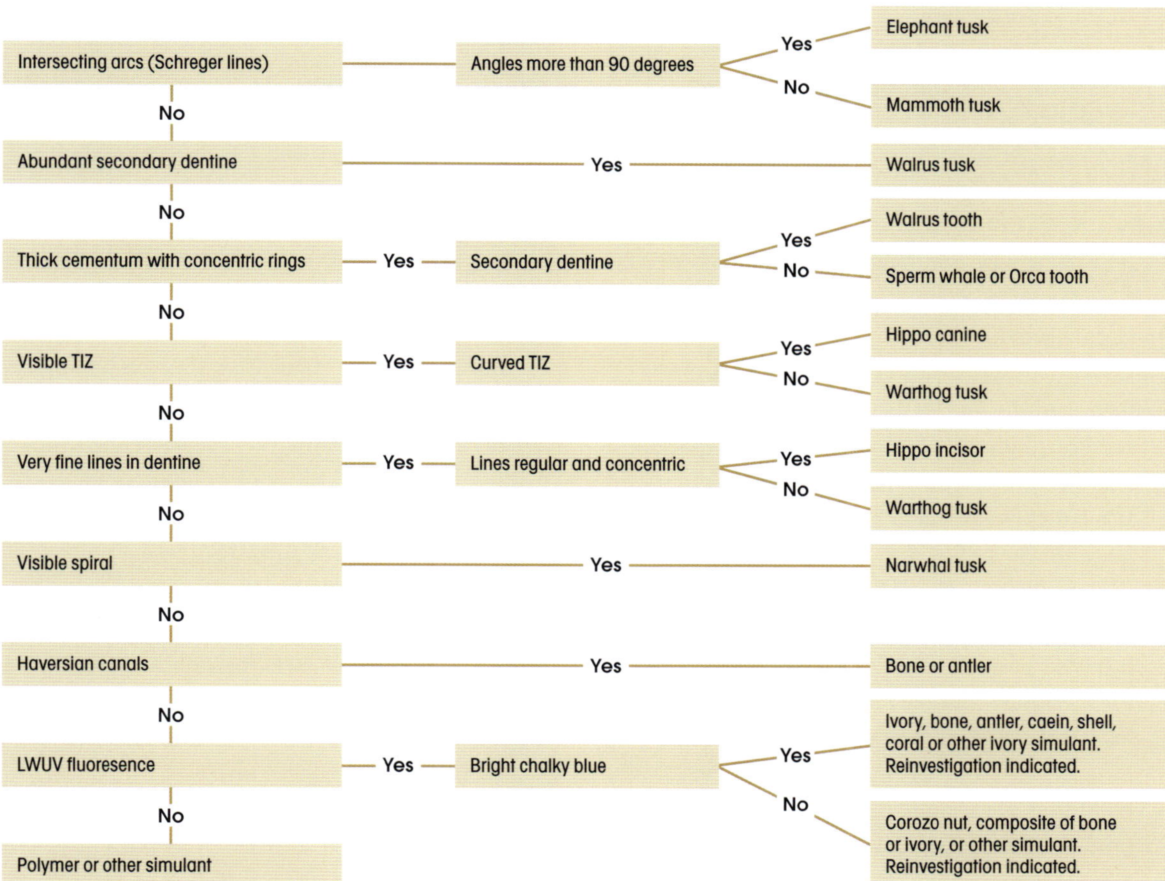

Glossary

10× magnifier A small hand lens which magnifies at ten-power. Much used in gemmology.

Al Andalus Islamic Iberia in mediaeval times, incorporating most of Spain, Portugal and Andorra, and part of France.

Aniline A liquid present in coal tar.

Aniline dye A synthetic dye incorporating aniline.

Aqua fortis Nitric acid.

Arabesque An abstract motif, based on flowering and scrolling plants. Common in Islamic art.

Axis A straight line about which a body can be perceived to rotate.

Baleen Sheets of keratinous material in the mouths of non-tooth-bearing whales, used for filtering food.

Bi-pedal Two-footed.

Black light American term for ultraviolet light.

Bodkin A blunt needle used for drawing tape or cord through fabric.

Cameo Small picture carved in low relief on a flat background, usually in a precious material.

Cancellous bone Type of bone with a solid outer edge and a light, honeycomb-looking interior.

Canines Pointed 'corner teeth' of most mammals, two each in upper and lower jaws.

Carbohydrate Group of organic compounds including sugar and starch.

Carbon 14 dating An advanced laboratory test used to discover the age of an item.

Carcass The hard core of an item.

Casque A horny protuberance on a bird's upper beak.

Cellulose The main constituent of plant cell walls and fibres.

Cementoblasts Cells in the jaw which produce cementum.

Cervid A member of the deer family.

CITES The Convention on International Trade in Endangered Species of Flora and Fauna. An international agreement among governments.

Collagen A protein. A main component of connective tissue, and the most abundant protein in mammals.

Cone-in-cone structure Growth pattern in teeth and tusks caused by a conical pulp laying down layers of dentine.

Cross-section A section perpendicular to the axis of the item. In a tusk (which does not have a straight axis), it refers to the axis at any one point along the length of the tusk.

Culling Organized killing of animals to prevent overpopulation.

Cusp The top of some forms of teeth.

Degrading process Deterioration and final collapse of a specimen.

Dentinal tubules Sub-microscopic canals in dentine.

Dentine Creamy-white material that makes up the bulk of most teeth or tusks.

Dentition The arrangement and number of teeth.

Destructive test A test that will in some way leave a mark on a specimen.

Diptych A work consisting of two panels, usually hinged together.

Distal Furthest from the jaw.

Ditty box Small box for odds and ends.

DNA Deoxyribonucleic acid, a material which is present in nearly all living organisms, and gives genetic information.

Electro-magnetic spectrum The entire range of electro-magnetic radiation, including X-ray and UV, IR, and visible lights.

Element The primary constituent of all matter.

Enamel The outer hard covering of some teeth or tusks.

Enzyme A catalyst causing a specific biochemical reaction.

Fake An object imitating another with the intention of deception.

Family See ***Taxonomic table***.

Fluorescence Normally refers to the visible light given out when a specimen is held in X-ray, UV, or some visible light (e.g. sunlight). Caused by the specimen absorbing energy in one wavelength and re-emitting it in another.

FTIR Fourier Transform Infrared Spectoscopy. An advanced laboratory test.

Hardness A term used in some sciences to denote how hard a material is when compared to other, known substances which are calibrated on a scale from 1–10.

Haversian canals Nutrient-bearing canals in bone.

Holy Sacrament Bread and wine, symbolizing the body and blood of Jesus Christ.

Hot point test A test used to determine the relative melting point of a material, and the smell emitted.

Hygroscopic The ability to absorb and re-emit moisture.

In the round A form of carving in which the item is carved all the way round, producing a three-dimensional figure.

Incisor A mammalian tooth, often situated between canines.

Inert Lacking chemical or physical reaction.

Intaglio Incised carving (as opposed to relief carving).

Inuk (pl. Inuit) An indigenous person of an Arctic region.

IUCN The International Union for Conservation of Nature. The oldest international conservation group.

IWC The International Whaling Commission. An organization tasked with the global conservation of whales.

Jagging wheel An instrument with a patterned wheel for crimping or cutting pastry. Often made by scrimshanders.

Keratin A protein very prevalent in the animal kingdom.

Keratinous Made of, or rich in keratin.

Longitudinal section Section taken along the length of an item, parallel to its axis.

Lustre The amount and quality of light reflected by a specimen.

LWUV, long wave ultraviolet light. See ***Ultraviolet light***.

Mandrake A plant of the nightshade family.

Midden An ancient rubbish tip.

Mohs' Scale The scale used to measure hardness, based on the ability of one material to scratch another one. Created in 1812 by Friedrich Mohs.

Molar A mammalian tooth at the back of the jaw.

Monochromatic light Light of one wavelength only.

Mordant A fixative used in staining or dyeing.

Morse A term sometimes used for walrus tusk ivory.

Musth The periodic state of heightened sexual sensitivity and aggressiveness in male elephants.

Natural material A material that is not man-made.

Nerve canal A narrow cavity containing a nerve.

Odontoblasts The cells in the pulp cavity of a tooth which produce dentine.

Oliphant An elephant tusk, carved and used as a hunting horn or for ceremonial purposes in mediaeval times in Al Andalus and Italy.

Opaque Not allowing light to pass through.

Owen Lines The name given to rings seen in proboscidean ivory when viewed in cross-section.

Oxodizing atmosphere An atmosphere in which there is oxygen present.

Patina The surface appearance acquired with age.

Pelagic Of the sea.

Peptide A chain of amino-acids.

Permafrost An area of ground or a region that is more or less permanently frozen.

Pinnipeds The family of animals that includes walruses and seals.

Plasticizer A substance that enables another substance to become malleable.

Pod A group of whales living together.

Polymer A large molecule, composed of many repeating units. Can be natural or synthetic, but is often loosely regarded as referring to a synthetic plastic.

Polyptych A work consisting of four or more panels joined together.

Poultice A substance applied to an object to draw out and absorb unwanted material.

Proboscidean Of the family encompassing elephants, mammoths and mastodons.

Protein A class of organic compounds containing nitrogen, which are an essential part of all living organisms.

Proximal Nearest to the jaw.

Pyxis (pl. pyxides) Small, lidded box, usually round in shape.

Raman Spectroscopy An advanced laboratory test, used to discover the identify of a sample.

Raw ivory Unworked and untreated ivory.

Reconstituted Dismantled and rejoined, often with an additive.

Reflected light Light bounced off the surface of an object.

Refractive index A measurement of the degree by which a material refracts (bends) visible light entering it from the surrounding air. A much-used gemmological test.

Reliquary A box for containing relics such as pieces of bone or clothing, usually purporting to be from a saint or holy figure.

Schreger Lines Intersecting arc pattern seen in proboscidean ivory when viewed in cross-section.

Scrivello A smaller-sized elephant tusk, often chosen for making billiard balls.

Semi-aquatic Living partly in water.

Semi-synthetic plastic A plastic consisting of part natural substances such as milk curds, plus chemicals.

Semi-translucent Allowing only some light to pass though.

Shagreen Leather made from the skin of a fish of the ray family.

Shaman A person regarded as having access to, and influence in, the spirit world, and able to practise divination and healing through going into a trance.

Simulant A material which simulates or copies a natural one, not necessarily with the intention of deception.

Species See *Taxonomic table*.

Specific gravity (SG) The ratio of the mass of a substance to the mass of an equal volume of water at a temperature of 4°C. A test much-used in gemmology.

Subsistence quota The quota of animals permitted to be killed by indigenous peoples as part of their cultural heritage and rights.

Sub-species See ***Taxonomic table***.

Suid A member of the pig family, including wild boar and warthog.

Synthetic Man-made substance or material.

Tabernacle (In the context of this book) refers to an ornamented cabinet which may contain the sacrament in churches, especially those of Catholic faith.

Taxonomic table A table classifying living organisms. It starts with Kingdom (e.g. 'animal'), and subdivides through Phylum, Class, Subclass, Order, Family, Genus and Species. The genus and species are the most used, and the genus is always written in italics.

Tines Points on antlers.

TIZ Tusk (or Tooth) Interstitial Zone. The space between convergent layers of dentine.

TRAFFIC The wildlife trade monitoring organization.

Translucent Permitting a little light to pass though.

Transmitted light Light which is passed through a specimen.

Triptych A work consisting of three panels, usually hinged together.

Tusker An animal with tusks. Most commonly used for bull elephants.

Ultraviolet light Light in the wavelengths beyond visible light, and which causes some materials (including ivory) to fluoresce. Measured in long wave (*LWUV*) and short wave (*SWUV*).

Unicorn A mythical animal bearing a horn protruding from its forehead.

UV Abbreviation for ultraviolet light.

Vitreous The name given to the lustre of an object when it appears glass-like.

Wavelength A measurement in the electromagnetic spectrum.

Bibliography

Alpers, Edward A. (1975), *Ivory and Slaves: the Changing Pattern of International Trade in East Central Africa to the Later Nineteenth Century*, University of California Press, Oakland, CA

Ashley, Clifford W. (renewed 1966), *The Yankee Whaler*, Dover Publications Inc., New York, NY

Baarsen, Reinier (2000), *17th Century Cabinets*, Rijksmuseum, Amsterdam

Bagnoli, Martina et al. (2010), *Treasures of Heaven: Saints, Relics and Devotion in Medieval, Europe*, The British Museum Press, London

Barley, Nigel (2010), *The Art of Benin*, The British Museum Press, London

Barnet, Peter et al. (1997), *Images in Ivory: Precious Objects of the Gothic Age*, Detroit Institute of Arts & Princeton University Press, Princeton, NJ

Barnet, Peter & Wu, Nancy (2005), *The Cloisters, Medieval Art and Architecture*, The Metropolitan Museum of Art, New York, NY

Barnett, Richard D. (1979), *Ancient Ivories in the Middle East*, Qedem, Monographs of the Institute of Archaeology, The Hebrew University of Jerusalem, Jerusalem

Benson, Peter (2010), *The Art of Carving Netsuke*, Guild of Master Craftsmen Publications, Lewes, East Sussex

Bordignon, Laura (2010), *The Golden Age of Japanese* Okimono, Antique Collectors' Club, Woodbridge, Suffolk

Brown, Grahame & Mann, William (2007), Elk Canines. An unusual source of ivory, *The Journal of the Gemmological Association of Hong Kong*, Vol. XXVIII, Hong Kong

Brown, Grahame (2009), Rare Ivories, *The Australian Gemmologist*, Vol. 23, No. 11a, Normanhurst, NSW

Burack, Benjamin (1984), *Ivory and Its Uses*, Charles E. Tuttle Co., North Clarendon, VT

Caldwell, David H. et al. (2011), *The Lewis Chessmen Unmasked*, NMS Publishing Ltd, Edinburgh

Canby, Sheila R. (2012,) *Islamic Art in Detail*, The British Museum Press, London

Clunas, Craig et al. (1984), *Chinese Ivories from the Shang to the Qing*, British Museum Publications, London

Collins, Matthew et al. (2010), ZooMS: the collagen barcode and fingerprints, *Spectroscopy Europe*, Vol. 22 No. 2, Chichester, West Sussex

Cook, Jill (2013), *Ice Age Art: Arrival of the Modern Mind*, The British Museum Press, London

Corbin, George A. (1988,) *Native Arts of North America, Africa, and the South Pacific: An Introduction*, Harper & Row, New York, NY

Crawford, Vaughn E. et al. (1980), *Assyrian Reliefs and Ivories in the Metropolitan Museum of Art*, The Metropolitan Museum of Art, New York, NY

Crill, Rosemary et al. (1999). *Arts of India 1550–1900*, V&A Publications, London

Cutler, Anthony (1985), *The Craft of Ivory*, Dumbarton Oaks, Washington DC

Dauphin, Y. & Williams, C.T. (2008), Chemical composition of enamel and dentine in modern reptile teeth, *Mineralogical Magazine*, Vol. 72(1), pp. 247–250.

Delort, Robert (1992), *The Life and Lore of the Elephant*, Thames & Hudson, London

DeMouthe, Jean Frances (2006), *Natural Materials: Sources, Properties and Uses*, Butterworth-Heinemann, Elsevier, Oxford

Dinger, Brigitte (1994), *Elfenbein Kreativer Mystischer Werkstoff*, Deutches Elfenbein Museum, Erbach

Espinoza, Edgard O. & Mann, Mary-Jacque (1992), *Identification Guide for Ivory and Ivory Substitutes*, WWF and CITES Secretariat, New York, NY

Ezra, Kate (1984), *African Ivories*, The Metropolitan Museum of Art, New York, NY

Glenn, Virginia (2003), *Romanesque and Gothic: Decorative Metalwork and Ivory Carvings in the Museum of Scotland*, NMS Publishing Ltd, Scotland

Gordon, Jonathan (1998), *Sperm Whales*, Colin Baxter Photography Ltd, Grantown-on-Spey, Scotland

Gouk, Penelope (1988), *The Ivory Sundials of Nuremberg 1500–1700*, Whipple Museum of the History of Science, Cambridge, England

Greep, S.J. et al. (1984), Archaeological Bone, Antler and Ivory, *United Kingdom Institute for Conservation, Occasional Papers* no. 5

Guérin, Sarah M. (2010), Avorio d'Ogni Ragione: The Supply of Elephant Ivory to Northern Europe in the Gothic Era, *Journal of Medieval History*, 36, pp. 156–74

Guérin, Sarah M. (2013), Forgotten Routes? Italy, Ifrïqiya and the Trans-Saharan Ivory Trade, *Al-Mas q*, Vol . 25, No. 1, pp. 70–91

Gulløv, Hans Christian (2005), *Grønlands For historie*, Gyldendal, Copenhagen

Hansen, Keld (1996), *Legetøj i Grønland*, Nationalmuseet, Copenhagen

Hofmeister, W. et al., Ivory and Species Conservation, proceedings of INCENTIVS Meetings 2004–2007.

Holtzapffel, Charles (1852), *Turning and Mechanical Manipulation ... on the Lathe, Vol. 1*, Holtzapffel & Co., London

Hutt, Julia (2012), *Japanese Netsuke*, V&A Publications, London

Ickowicz, Pierre (2008), *La vie du Christ*, Les cahiers de l'ivoire, Château-Museé de Dieppe

Ickowicz, Pierre (2010) *Les Medaillons en ivoire découpé*, Les cahiers de l'ivoire, Château-Museé de Dieppe

Jaffer, Amin (2002,) *Luxury Goods from India: The Art of the Cabinet-Maker*, V&A Publications, London

Jülich, Theo (2007), *Die Mittelalterlichen Elfenbeinarbeiten des Hessischen Landesmuseums Darmstadt*, Schnell & Steiner GmbH, Regensburg

Kinsey, Robert O. (1991), *Ojime: Magical Jewels of Japan*, Harry N. Abrams Inc., New York, NY

Knudtson, Peter (1998), *The World of the Walrus*, Sierra Club Books, San Francisco, CA

Liang, Jie; Li, Haibo; et al. (2014), Identification Characteristics of Natural and Imitation Hornbill Ivory, *The Journal of Gemmology*, 34(1), pp42 –9

Liebgott, Niels-Knud (1985), *Elfenben*, Nationalmuseet, Copenhagen

Lister, Adrain, & Bahn, Paul (2007), *Mammoths: Giants of the Ice Age*, University of California Press, Oakland, CA

Locke, Michael, Structure of Ivory (2008), *Journal of Morphology 269*, pp. 423–450

Mann, William R. & Marts, Charles M. (2013), *Ivory Identification: A Photographic Companion.* CreateSpace Publishing Platform

McArthur, Meher (2005), *The Arts of Asia*, Thames & Hudson, London

McManus, Michael (1997), *A Treasury of American Scrimshaw: A Collection of the Useful and Decorative.* Penguin Books, New York, NY

Miles, A.E.W. and White, J.W. (1960), Ivory, Meeting at the Royal College of Surgeons, London.

O'Donoghue, Michael (ed.) (2006), *Gems* (6th edn), Butterworth-Heinemann, Elsevier, Oxford

Oxford Radiocarbon Accelerator Unit (2011), Radiocarbon Web-Info., *Oxford Radiocarbon Accelerator Unit and Waikata Radiocarbon Dating Laboratory*

Paine, Stefani (1995), *The Nature of Arctic Whales*, Greystone Books, Vancouver, BC

Parker, Elizabeth C., & Little, Charles T. (1994), *The Cloisters Cross, Its Art and Meaning*, The Metropolitan Museum of Art, New York, NY

Pedersen, Alwin (1951), *Rosmarus, en beretning om hvalrossens liv og historie,* Gyldendal, Copenhagen

Penniman, T.K. (1984), *Pictures of Ivory and other Animal Teeth, Bone and Antler*, Oxford University Press, Oxford

Pratt, Henry Sheldon (1901), *Monograph on Ivory Carving in Burma,* Facsimile, Kessinger Legacy Reprints

Prickett, Nigel (2004), *Maori Origins, from Asia to Aotearoa*, Auckland Museum, Auckland

Quye, Anita & Williamson, Colin (1999), *Plastics Collecting and Conserving*, NMS Publishing Ltd, Edinburgh

Rabenheimer, E.J.; Bosman, M.C., et al. (1998), Histogenesis of the chequered pattern of ivory of the African elephant, *Archives of Oral Biology* 43, pp. 969–977

Randall Jr., Richard H. (1985), *Masterpieces of Ivory from the Walters Art Gallery*, Hudson Hills Press, New York, NY

Raubenheimer, E.J. (2000), Early Development of the Tush and the Tusk of the African Elephant, *Archives of Oral Biology,* 45, pp. 983–6

Raubenheimer, E.J.; van Heerden, W.F.P., et al. (1995), Morphology of the Deciduous Tusk (tush) of the African Elephant, *Archives of Oral Biology*, 40(6), pp. 571–6

Ritchie, C.I.A (1969), *Ivory Carving*, Arthur Barker Ltd, London

Roesdahl, Else (1995), *Hvalrostand elfenben og nordboerne I Grønland*, Odense Universitets-forlag

Rosing, Jens (1986), *Havets enhjørning*, Wormianum, Denmark

Ross, Doran H. (ed.) (1992), *Elephant. The Animal and Its Ivory in African Culture*, Fowler Museum of Cultural History, California

Rosser-Owen, Mariam (2004), *Ivory,* National Council for Culture, Arts and Heritage, Doha

Rosser-Owen, Mariam (June 2013), West African Trade and the Ivories of Al Andalus, Conference: 'Ivory trade and exchange in Late Antiquity and Early Islam', Warburg Institute

Schabillion, Shiri (1989), *All in a Nutshell: The Story of the Vegetable Ivory Nut*, The Mississippi Petrified Forest, MS

Shalem, Avinoam (2004), *The Oliphant*, Koninklijke Brill N.V. Leiden

Sheriff, Abdul (1987), *Slaves, Spices and Ivory in Zanzibar*, Ohio University Press, OH

Siddons, George A. (1837), *The Cabinet Maker's Guide*, Facsimile, Lightning Source, Milton Keynes

Sims, Margaret E. (2010), Unusual Appearance of Schreger-like Pattern in *Hippopotamus Amphibious* Ivory: Wildlife Forensics Investigation of a *Netsuke*, *Forensic Science International 200*, e19–e20

Soxhlet, V.H., Morris, et al. (1902), *The Art of Dyeing and Staining Marble, Bone, Horn, Ivory, etc.*, Scott, Greenwood & Sons, London

Simpson, Dr. St John (2011), *Indian Ivories from Afghanistan: The Begram Hoard*, The British Museum Press, London

Trusted, Marjorie (ed.) (2007), *The Making of Sculpture*, V&A Publications, London

Trusted, Marjorie (2013), *Baroque and Later Ivories*, V&A Publications, London

Vickers, Michael et al. (1987), *Ivory: A History and Collector's Guide*, Thames & Hudson, London

von Folsach, Kjeld & Meyer, Joachim (eds) (2005), *Journal of the David Collection*, Vol. 2, No. 1, The David Collection, Copenhagen

von Folsach, Kjeld & Meyer, Joachim (eds) (2005) *Journal of the David Collection*, Vol. 2, No. 2, The David Collection, Copenhagen

Walker, John Frederick (2009), *Ivory's Ghosts: The White Gold of History and the Fate of Elephants*, Atlantic Monthly Press

Wardwell, Allen (1986), *Ancient Eskimo Ivories of the Bering Strait*, Hudson Hills Press, New York, NY

West, Janet & Credland, Arthur G. (1995), *Scrimshaw: The Art of the Whaler*, Hutton Press Ltd

Wilk, Christopher (1996), *Western Furniture 1350 to the Present Day*, Philip Wilson, and V&A Publications, London

Williamson, G.C. (1938), *The Book of Ivory*, Latimer Trend & Co Ltd, Plymouth, Devon

Williamson, Paul (2010), *Medieval Ivory Carvings: Early Christian to Romanesque*, V&A Publications, London

Wilson, Derek & Ayerst, Peter (1976), *White Gold: The Story of African Ivory*, Heinemann, London

Index

Illustrations are indicated in **bold**